Social and Organizational Performance Review
Concepts and Research

Review of the Performance Improvement Institute
PhD and MBA program on Social, Organizational
and Individual Performance & Value Creation
at

The Sonora Institute of Technology
ITSON

October, 2009

Year 1 – Volume 1

Social and Organizational Performance Review

AuthorHouse™
1663 Liberty Drive
Bloomington, IN 47403
www.authorhouse.com
Phone: 1-800-839-8640

First published by AuthorHouse 11/6/2009

ISBN: 978-1-4490-3057-5 (sc)

Library of Congress Control Number: 2009911951

Printed in the United States of America
Bloomington, Indiana

This book is printed on acid-free paper.

authorHOUSE®

Foreword

We can be the masters of change or the victims of it. The Sonora Institute of Technology (ITSON) has chosen a leadership role in creating a university that focuses on adding measurable value to all stakeholders, including ecosystems that provide the vehicles for planned, societal-focused change. As part of the ITSON commitment, the Performance Improvement Institute (PII) has been created—an institute that includes the performance of society, organizations, and people and aligns them to achieve ethical and measurable value.

There is an incubator dimension to the PII where now more than 20 organizations are using a Mega—societal value added—focus. Energizing students, faculty, and community resources, these organizations are applying the practical, ethical, and proven techniques of Mega thinking and planning and measuring both their intentions and results on a societal bottom line linked with a conventional bottom line. The intention is to add societal value using their organizations as the vehicle. This is very pragmatic.

In addition to the PII sponsoring organizations, the PII is in the second phase of a PhD and an MBA in Mega thinking, planning, and doing. The graduates of these programs will be the faculty and business leaders of the future, not just in Mexico but also in our shared world.

Now, as part of the PII and its commitment to results-based and research-based performance accomplishment, this is the first issue of Social and Organizational Performance Review (SOPR). It includes timely work by ITSON faculty, staff, and others committed to adding societal value. This first issue includes the concepts and tools of Mega thinking and planning, and proven tools and techniques for designing, delivering, and evaluation organizational and societal impact.

While some of the work that appears in this and future issues might have been also published elsewhere, they are provided here with full attribution of initial source. We also will be publishing new work and the results of the applied research and development of ITSON students, their sponsors, and sponsoring organizations. We will also find useful work done throughout the world and bring it to you through the SOPR.

The most important part of this Review is the reader.

We invite you to suggest contributions, provide feedback on what we publish, and become a partner in measurably improving our shared world. Come visit our facility at the ITSON Campus and review our capabilities and progress at www.piionline.org and www.piiblog.com

We invite you to join this adventure to truly change our shared world for the better.

Ingrid-Guerra Lopez
Lead Editor
(iguerra@wayne.edu)

Mariano Bernardez
Executive Editor
(mbernardez@expert2business.com)

Roger Kaufman
Contributing Editor
(rkaufman@nettally.com)

Social and Organizational Performance Review

Table of Contents

Foreword   3

Introduction   7

Abstracts   9

Mega Thinking and Planning: an introduction to Defining and
Delivering Individual and Organizational Success   15

By

Roger Kaufman, CPT, PhD

Using technology and innovation for the planning social and economic
transformation in a region of Mexico   35

By

Gonzalo Rodriguez Villanueva & Roberto Lagarda Silva

Minding the business of business: tools and models to design and
measure wealth creation   73

By

Mariano L. Bernardez, PhD., CPT

Surviving Performance Improvement "Solutions": Aligning
Performance Improvement Interventions   125

By

Mariano Bernardez, PhD., CPT

Are Performance Improvement Professionals Measurably Improving
Performance? A look at what PIJ and PIQ have to say about the
current use of evaluation and measurement in the field of performance
improvement   155

By

Ingrid Guerra-Lopez, PhD. & Hillary N. Leigh

Social Responsibility of a Profession: An Analysis of Faculty
Perception of Social Responsibility Factors and Integration into
Graduate Programs of Educational Technology   179

By

Stephanie L. Moore, PhD.

Introduction

Welcome to the first issue of *Social and Organizational Performance Review: Concepts and Research.*

This new journal is a reflection of the work that is being conducted at the Institute for Social and Organizational Performance in Sonora, Mexico, which illustrates how organizations can achieve social and organizational transformation using rigorous performance improvement concepts and methodologies.

This issue includes foundational concepts, specific case studies, and research relevant to rigorous performance improvement approaches.

In the first article, Roger Kaufman reminds us that all organizations are means to societal ends, and shares some conceptual and practical tools for Mega Planning, including the Organizational Elements Model (OEM), the Six Critical Success Factors, and the Six-Step Problem Solving model.

Next, Rodriguez and Lagarda show us that it is possible to achieve social transformation and accelerate the economic growth and social development of a region by using technology and innovation based on the shared vision and efforts of a university, government, and the private sector.

In the next article, Mariano Bernardez describes a rational for developing successful new business based on the notion that the business of business is after all to continually add value to all stakeholder, including making clients successful so that they can continue to purchase and recommend its products and services.

In his next article, Mariano Bernardez illustrates how to avoid the negative effects of poorly coordinated performance improvement interventions by introducing a systemic, multi-level framework to align performance improvement interventions, avoid systemic disruption, measure and eliminate over costs, rework and negative side effects of change.

Then, Guerra-López and Leigh discuss the central role of measurement and evaluation in ensuring the credibility of performance improvement professionals and demonstrating the value they add to clients. The authors present the findings of a content analysis conducted with ten years of PIJ and PIQ articles to further explore this issue.

To conclude this special issue, Stephanie Moore shares the findings from an investigation on deriving an empirical definition of social responsibility, focused on faculty perceptions of whether a set of ethical outcomes are relevant to the performance improvement and educational technology fields.

We trust this first issue of *Social and Organizational Performance Review: Concepts and Research* will be useful for anyone who is serious about measurably and reliably improving performance at all levels of the organization and society.

Ingrid Guerra-López
Lead Editor

Abstracts

Mega Thinking and Planning: an introduction to Defining and Delivering Individual and Organizational Success
By
Roger Kaufman, CPT, PhD

All organizations are means to societal ends, and thus Mega thinking and planning starts with a primary focus on adding value for all stakeholders, including our shared society.

It is pragmatic, realistic, practical, and ethical.

Defining and achieving continual organizational success is possible.

It relies on three basic elements

(1) *A societal value-added "frame of mind"*: your perspective and commitment about your organization, people, and our shared world

(2) *A shared determination and agreement on where to head and why:* all people who can and might be impacted by the shared objectives must agree on purposes and results criteria, and

(3) *Pragmatic and basic tools.* Following are the basic concepts for thinking and planning Mega in order to define and deliver value to internal and external partners; defining and delivering individual and organizational success.

Using technology and innovation for the planning social and economic transformation in a region of Mexico
By
Gonzalo Rodriguez Villanueva & Roberto Lagarda Silva

This article shows the possibility of achieving social transformation using technology and innovation through the design and implementation of a plan in order to accelerate the rate of economic growth and social development of a region.

The model includes three basic actors: university, government, and the private sector in order to create a regional system of innovation.

These three allies must shared the vision and have participated during the entire process mainly with five elements:

(a) Defining an strategic plan for regional development based on innovation;

(b) Supporting the creation of ecosystems and productive chains;

(c) Prioritizing investment in technology of information, communication and transport in order to improve territorial integration;

(d) Creating a safe and healthy environment to attract and retain investment; and

(e) Developing a program based on a state policy and administered by strategic projects.

Minding the business of business: tools and models to design and measure wealth creation

By

Mariano L. Bernardez, PhD., CPT

What is the business of business? How can planners and investors anticipate the true chances of failure and success of a business idea?

This article describes a rationale for developing successful new business based on a simple, sensible idea: the business of any business is to make its clients successful enough as to continue purchasing and recommending its products and services and continue to add value to all stakeholders.

Using a double-bottom line, triple top line business case and a wealth creation flowchart, the author shows how to demonstrate measurable benefit as the core of a business proposition and engineer the creation and delivery of value based on a carefully designed client experience.

This new technology, illustrated with examples of multiple business incubated at the Sonora Institute of Technology (ITSON), combines Roger Kaufman's *Organizational Elements Model* –OEM- with Dale Brethower's and Geary Rummler's *Anatomy Of Performance* –AOP- models in a simple, straightforward process and is supported by an extensive research bibliography.

Surviving Performance Improvement "Solutions": Aligning Performance Improvement Interventions

By

Mariano L. Bernardez, PhD., CPT

How can organizations avoid the negative, sometimes chaotic effects of multiple, poorly coordinated "performance improvement" interventions? How can we avoid punishing our external clients or staff with the "side effects" of "solutions" that might benefit our bottom line or internal efficiency at the expense of the value received or perceived by clients and investors and our shared world, a world we all live and depend upon?

Facing multi-billion dollar consulting industry pushing every year new "solutions" that might end causing new and unexpected problems, serious performance consultants and managers know that blaming the "law of unintended consequences" will not prevent clients from leaving, staff turnover, general organizational turmoil, and even national and international consequences associated with change.

Since change is also vital, this article introduces a systemic, multi-level framework to align performance improvement interventions, avoid systemic disruption, measure and eliminate over costs, rework and negative side effects of change.

Are Performance Improvement Professionals Measurably Improving Performance? A look at what PIJ and PIQ have to say about the current use of evaluation and measurement in the field of performance improvement

By

Ingrid Guerra-Lopez, PhD. & Hillary N. Leigh

Measurement and evaluation are at the core of reliably improving performance.

It is through these central mechanisms that performance improvement professionals are able to demonstrate the true worth of their efforts.

However, the true value of the contributions performance improvement practitioners make is inconclusive.

This article presents a content analysis of ten years worth of PIJ and PIQ articles as an initial data point to be used for professional reflection and further exploration into the intentions and practices of performance improvement practitioners.

Social Responsibility of a Profession: An Analysis of Faculty Perception of Social Responsibility Factors and Integration into Graduate Programs of Educational Technology

By
Stephanie L. Moore

Although ethics are commonly regarded as an important characteristic and performance attribute, they are also regarded as a slippery or ill-defined topic leaving practitioners and faculty flat-footed in how to teach and assess it.

This article reports part of the findings from an investigation on deriving an empirical definition of ethics, namely social responsibility, focused on faculty perceptions of whether a set of ethical outcomes identified as "social responsibility" are relevant to the profession of educational technology/human performance technology and therefore ought to be a part of graduate curricula.

Findings focus specifically on faculty perceptions of the role of the profession educational technology/human performance technology in society and its impact on society.

While other professions have increasingly identified their responsibilities to society and integrated content into curricula, early findings indicate a lag in the instructional design/human performance technology community.

Mega Thinking and Planning: an introduction to Defining and Delivering Individual and Organizational Success

By

Roger Kaufman, CPT, PhD

A Societal Value-Added Perspective and Frame of Mind

Adding value to our shared society, using your organization as the primary vehicle is Mega thinking and planning.

It is straight-forward, and sensible. From this shared societal value-added frame, everything one uses, does, produces, and delivers is linked to deliver shared and agreed-upon positive organizational as well as societal results.

This societal frame of reference, or paradigm, I call the *Mega* level of thinking and planning. If you are not adding value to our shared society you have no assurance that you are not subtracting value. Starting with Mega as the central focus is strategic thinking and provides the data based for strategic planning.

A central question that every organization should ask and answer is:

If Your Organization is the Solution, What's the Problem?

This fundamental proposition—using a Mega focus—represents a shift from the usual attention only on oneself, individual performance improvement, and one's organization to making certain you also add value to external clients and society.

An Overview of the Basic Concepts and Tools for Mega Thinking and Planning

There are three basic guides, or templates, that help define and achieve individual and organizational success. Each is provided in much greater detail in several books (see the references), but for our entry into Mega Planning and strategic thinking, following is the short introduction to these three guides.

Guide One: The Organizational Elements Model (OEM)

It is important to define and link (align) what any organization uses, does, produces, and delivers to achieve external client and societal value added.

A tool for making sure that everything an organization, uses, does, produces, and delivers does add value to external clients and society is called the *Organizational Elements Model (OEM)* and is shown in Table 1.[1]

For each Element, there is an associated level of planning: *strategic planning* (and thinking) starts with Mega while *tactical planning* starts with Macro and *operational planning* at Micro.

These elements are also useful for defining the basic questions every organization must ask and answer as provided in Figure 3.

Name of the Organizational Element	Name of the Level of Planning and Focus	Brief Description	Type of Planning

---

[1] It should be noted that the OEM is useful for making sure there is inclusion of each factor in organizational success; it does not actually do the alignment.

Outcomes	Mega	Results and their consequences for external clients and society (shared Ideal Vision)	Strategic[2]
Outputs	Macro	The results an organization can or does deliver outside of itself	Tactical
Products	Micro	The building block results that are produced within the organization	Operational
Processes	Process	The ways, means, activities, procedures, projects, methods used internally	
Inputs	Input	The human, physical, financial resources an organization can or does use	

---

[2] These definitions of Strategic and Tactical are different from other conventional usage. I suggest that defining "strategic" as adding value to society and "tactical" as defining the best ways and means to achieve societal results is more pragmatic and encourages planners to justify any organizational mission in terms of Mega.

Table 1. The five levels of results, the levels of planning, and a brief description.

These elements are also useful for defining the basic questions every organization must ask and answer as provided in Figure 3.

Guide Two: Six Critical Success Factors

Following are what provides an essential framework of this approach and for Mega planning. Unlike conventional "critical success factors," these are factors for successful planning, not just for the things that an organization must get done to meet its mission. These are for Mega planning, regardless of the organization.

The Six Critical Success Factors for Mega thinking and planning are shown in Figure 1.

FIGURE 1. THE SIX CRITICAL FACTORS FOR MEGA THINKING AND PLANNING

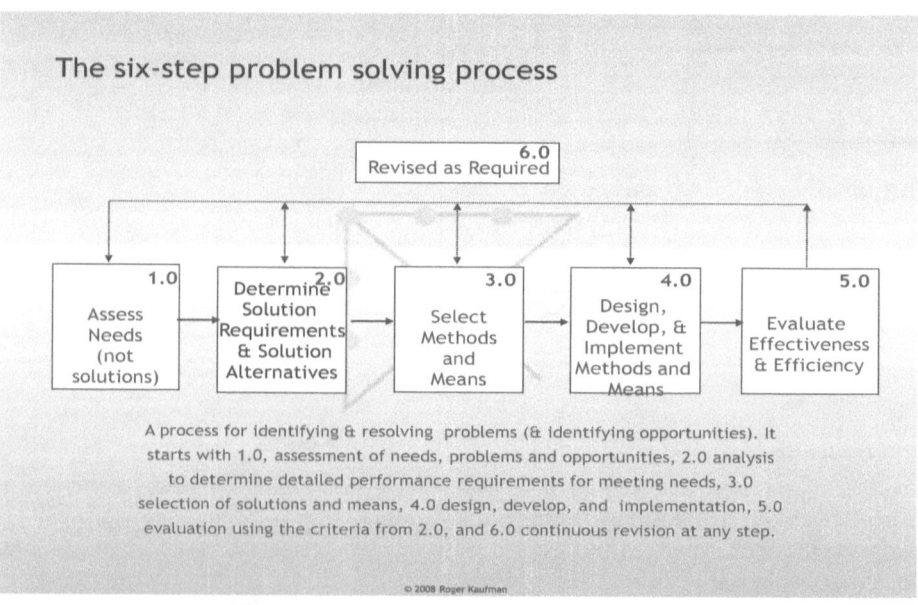

The six-step problem solving process

A process for identifying & resolving problems (& identifying opportunities). It starts with 1.0, assessment of needs, problems and opportunities, 2.0 analysis to determine detailed performance requirements for meeting needs, 3.0 selection of solutions and means, 4.0 design, develop, and implementation, 5.0 evaluation using the criteria from 2.0, and 6.0 continuous revision at any step.

© 2008 Roger Kaufman

Unlike many other presentations of critical success factors, these relate to any organization public or private. Most "critical success factors" discussed in the management literature refer to organization-specific factors related to their unique business. These apply to any organization and are "above" any organizational-specific factors.

Guide Three: A six-step problem solving model,

A process for getting from *What Is* to *What Should Be* is shown in Figure 2.
c
These functions include:

(1.0) identifying problems based on needs,
(2.0) determining detailed solution requirements and identifying (but not yet selecting) solution alternatives,
(3.0) Selecting solutions from among alternatives,
(4.0) Implementation,
(5.0) Evaluation and
(6.0) Continuous improvement (at each and every step):

FIGURE 2. THE SIX-STEP PROBLEM SOLVING PROCESS: A PROCESS FOR IDENTIFYING AND RESOLVING PROBLEMS AND IDENTIFYING OPPORTUNITIES (Adapted from Kaufman(1992, 1998, 2000, 2006a, b,)

Each time you want to identify opportunities and resolve problems systematically get from current results and consequences to desired ones, use the six-step process. This Six-step process In Figure 1 (Based in part on Kaufman, 1992, 1998, 2000, 2006) below allows the identification of opportunities before immediately moving to solve problems. The Mega thinking and planning approach does not assume that improving performance with the existing situation is automatically useful. Often, an organization can improve performance only to later discover that the performance in question does not add measurable value to the organization or to the shared society.

To be successful—to do and apply Mega Planning—you have to realize that yesterday's methods and results often are not appropriate for tomorrow. Most planning experts agree that the past is only prologue, and tomorrow must be crafted through new patterns of perspectives, tools, and results. The tools and concepts for meeting the new realities of society, organizations, and people are linked to each of the Six Critical Success Factors.

The details and how-to's for each of the three guides are also provided in the referenced sources. The three basic "guides" or templates should be considered as forming an integrated set of tools—like a fabric—instead of only each one on their own.

A Mega Planning framework has three phases:
- Scoping,
- Planning, and
- Implementation/Continual Improvement.

During the Scoping phase, one may find opportunities that were not readily apparent from most reactive problem solving approaches to strategic planning. From this framework, specific tools and methods are provided to do Mega Planning. It is not complex, really. If you simply use the three guides you will be able to put it all together.

When doing Mega planning, you and your associates will ask and answer the following questions shown in Figure 3. This also identifies the Organizational Elements in terms of the questions you and your organization should (and must) ask and answer:

FIGURE 3. THE BASIC QUESTIONS EVERY ORGANIZATION MUST ASK AND ANSWER (BASED ON KAUFMAN 2006)

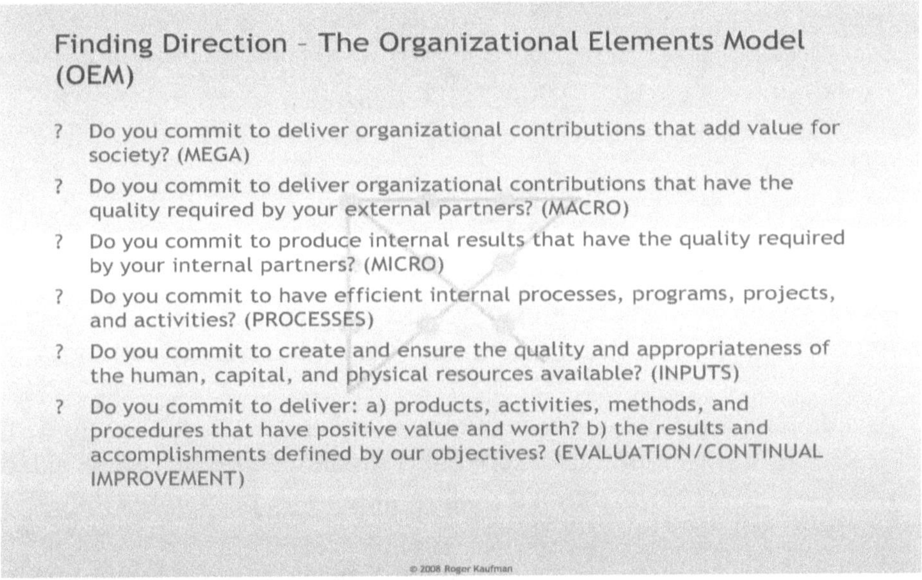

Finding Direction – The Organizational Elements Model (OEM)

? Do you commit to deliver organizational contributions that add value for society? (MEGA)

? Do you commit to deliver organizational contributions that have the quality required by your external partners? (MACRO)

? Do you commit to produce internal results that have the quality required by your internal partners? (MICRO)

? Do you commit to have efficient internal processes, programs, projects, and activities? (PROCESSES)

? Do you commit to create and ensure the quality and appropriateness of the human, capital, and physical resources available? (INPUTS)

? Do you commit to deliver: a) products, activities, methods, and procedures that have positive value and worth? b) the results and accomplishments defined by our objectives? (EVALUATION/CONTINUAL IMPROVEMENT)

© 2008 Roger Kaufman

A "yes" to all questions will deliver Mega planning and allow you to prove that you have added value. . . something that is becoming increasingly important. These questions relate to Guide One that defines each organizational element in terms of its label and the question each addresses. If you use and do all of these you will align everything you use, do, produce, and deliver to adding measurable value to yourself, your organization, and to external clients and society.

Mega planning is proactive. Many approaches to organizational improvement wait for problems to happen and then hastily respond. But there is a temptation to react to problems and never take the time to plan so surprises are fewer and success is defined—before problems spring up—and then systematically achieved.

The Six Critical Success Factors in Brief

Examining each of the Six Critical Success Factors –Guide Three --to get a sense for the frame of mind (or paradigm) Mega planning provides.

Critical Success Factor 1. Don't assume that worked before will work now.

Don't assume that which worked for you and others in the past will work in the future.

There is evidence just about everywhere we look that tomorrow is not a linear projection—a straight-line function—of yesterday and today.

Examples include car manufacturers that squander their dominant client base by shoving unacceptable vehicles into the market and airlines that focus on shareholder value and ignore customer value.

An increasing number of credible authors have been, and continue to tell us that the past is, at best, prologue and not a harbinger of what the future will be.

In fact, old paradigms can be so deceptive that Tom Peters (1997) suggests that "organizational forgetting" must become conventional organizational culture.

Times have changed, and anyone who doesn't also change appropriately is risking failure. It is vital to use new and wider boundaries for thinking, planning, doing, and delivering.

Doing so will require getting out of current comfort zones.

Not doing so will likely deliver failure.

Critical Success Factor 2: Use an Ideal Vision (Mega) as the underlying basis for all strategic think, planning, and doing (Don't Be Limited to Your Own Organization)

Here is another area that requires some change from the conventional ways of doing planning. This Ideal Vision is identical for all organizations, public and private. One planning for one's organization simply identifies which of the variables they commit to deliver and move ever-closer toward.

An Ideal Vision, Exhibit 1, identifies the kind of world we want to help create for tomorrow's child. It identifies measurable variables that can be used to (1) identify needs at the Mega/societal level, (2) provide measurable criteria for an organization's mission, and (3) assure that everything an organization uses, does, produces, and delivers will add measurable value to all stakeholders.

EXHIBIT 1. THE IDEAL VISION (KAUFMAN, 2006)

From this societal-linked Ideal Vision, each organization can identify what part or parts of the Ideal Vision they commit to deliver and move ever-closer toward. If we base all planning and doing anchored on the Ideal Vision of the kind of society we want for future generation, we can achieve "strategic alignment" for what we use, do, produce, deliver, and the external payoffs for our Outputs.

Basic Ideal Vision:
The world we want to help create for tomorrow's child

- There will be no losses of life nor elimination or reduction of levels of well-being, survival, self-sufficiency, quality of life, from any source including (but not limited to):
    - war and/or riot and/or terrorism
    - unintended human-caused changes to the environment including permanent destruction of the environment and/or rendering it non-renewable
    - murder, rape, or crimes of violence, robbery, or destruction to property
    - substance abuse
    - disease
    - starvation and/or malnutrition
    - destructive behavior (including child, partner, spouse, self, elder, others)
    - accidents, including transportation, home, and business/workplace
    - discrimination based on irrelevant variables including color, race, age, creed, gender, religion, wealth, national origin, or location
- Poverty will not exist, and every woman and man will earn as least as much as it costs them to live unless they are progressing toward being self-sufficient and self-reliant. No adult will be under the care, custody or control of another person, agency, or substance: all adult citizens will be self-sufficient and self-reliant as minimally indicated by their consumption being equal to or less than their production.

© 2008 Roger Kaufman

The Ideal/Mega Vision is not the same as design and development but simply provides a "North Star" toward which everyone in the organization can develop their products and steer closer toward. On very simple decision criteria and time a decision to be made is objectively ask and answer *will this take us closer or further away from Mega?*

Mega thinking and planning is about defining a shared success, achieving it, and being able to prove it. Mega thinking and planning is a focus not on one's organization alone but upon society now and in the future.

It is about adding measurable value to all stakeholders. (Mega thinking and planning is not a tool for the actual design, development, implementation, and evaluation of organizational effectiveness but rather for scoping and setting requirements and for checking on measurable contributions and alignment.

The operational analysis, design, development, implementation, and evaluation/continual improvement is best done by a number of excellent models and approaches, such as Bernardez, 2006a,2006b; Brethower, 2006; Gilbert, 1978; Guerra, 2003; Kaufman, R., Guerra, I., and Platt, W. A. ,2006; Rummler 2004; Watkins, 2007, among others.)

Mega thinking and planning has been offered for many years, perhaps first formally with Kaufman, 1972 and further developed in Kaufman & English, 1979, and continuing through this article. In one form another, using a societal frame for planning and doing has shown up in the works of other respected thinkers, including Senge (1990) and more recently Prahalad (2005) and Davis (2005). And this concept was introduced by Kaufman, Corrigan, & Johnson,1969 .

Appropriately, there seems to be a lessening of resistance to Mega thinking and planning; there continues a migration from individual performance as the preferred unit of analysis for performance improvement to one that includes a first consideration of society and external stakeholders; It is responsible, responsive, and ethical to add value to all.

Critical Success Factor 3. Differentiate between Ends and Means

Focus on "what" (mega/outcomes, macro/outputs, micro/products) before "how." People are "doing-types." We want to swing right into action and in so doing we usually jump right into solutions (means) before we know the results (ends) we must deliver. Writing and using measurable performance objectives is something upon which almost all performance improvement authors agree. Objectives correctly focus on ends and not methods, means, or resources.

Ends—"What"—sensibly should be identified and defined before we select "How" to get from where we are to our destinations. If we don't select our solutions, methods, resources, and interventions on the basis of what results we are to achieve, what do we have in mind to make the selections of means, resources, or activities?

Focusing on means, processes, and activities is usually more comfortable as a starting place for conventional performance improvement initiatives.

Starting with means, for any organization and performance improvement initiative, would be as if you were provided process tools and techniques without a clear map that included a definite destination identified (along with a statement of why you want to get to the destination in the first place).

Also, a risk for starting a performance improvement journey with means and processes would be the fact that there would be no way of knowing whether your trip is taking you toward a useful destination or the criteria for telling you if you were making progress.

It is vital that successful planning focuses first on results (and not "how")—useful performance in measurable terms—for setting its purposes, measuring progress and providing continuous improvement toward the important results, and for determining what to keep, what to fix, and what to abandon.

This rigorous base sets the stage for another related Critical Success Factor 3 (Use and Link all Three Levels of Results) through application of the Organizational Elements Model (OEM) and for Critical Success Factor 4 (Prepare objectives that have indicators of how you will know when you have arrived). The OEM relies on a results-focus because it defines what every organization uses, does, produces, delivers, and the consequences of that for external clients and society.

Critical Success Factor 4: Prepare objectives—including those for the Ideal Vision (Mega) and the mission that have rigorous indicators to tell if you have arrived at your intended destination.

It is vital to state, precisely, measurable, and rigorously, where you are headed and how to tell when you have arrived.[3]

Statements of objectives must be in performance terms so that one can plan how best to get there, how to measure progress toward it. And everything is measurable, in spite of conventional wisdom, so don't deceive yourself into thinking you can dismiss important results as being "intangible" or "non-measurable." [4] Increasingly organizations throughout the world are increasingly focusing on Mega-level results.[5]

Objectives, at all levels of planning, activity, and results, are absolutely vital. And everything is measurable, so don't kid yourself into thinking you can dismiss important results as being "intangible" or "non-measurable."

It is only sensible and rational to make a commitment to measurable purposes and destinations.
Organizations throughout the world are increasingly focusing on Mega-level results

---

[3] An important contribution of strategic planning at the Mega level is that objectives can be linked to justifiable purpose. Not only should one have objectives that state "where you are headed and how you will know when you have arrived," they should also be justified on the basis of "why you want to get to where you are headed." While it is true that objectives only deal with measurable destinations, useful strategic planning adds the reasons why objectives should be attained.

[4] There are four scales of measurement: nominal, ordinal, interval, and ratio. If you can't name it, how do you know it even exists?

[5] Cf. Kaufman, Watkins, Triner, & Stith, 1998:Summer, and Davis, 2005.

Critical Success Factor 5: Use and Align all three levels of Planning and Results.

As we noted in Critical Success Factor 2, it is vital to prepare all objectives that focus only on ends; never just on means or resources.

There are three levels of results, shown in Table 2, that are important to target and link.

There are three levels of planning and results, based on who is to be the primary client and beneficiary of what gets planned, designed, and delivered. For each level of planning there are three associated levels of results (Mega/Outcomes, Macro/Outputs, Micro/Products).

Table 2. The levels of planning and results that should be linked during planning, doing, and evaluation and continuous improvement and there are three levels of planning.

PRIMARY CLIENT AND BENEFICIARY	NAME FOR THE LEVEL OF PLANNING	NAME FOR THE LEVEL OF RESULT	TYPE OF PLANNING
Society and External Clients	Mega	Outcomes	Strategic
The Organization Itself	Macro	Outputs	Tactical
Individuals and Small Groups	Micro	Products	Operational

Critical Success Factor 6: Define "need" as a gap between current and desired results (Not as Insufficient Levels of Resources, Means, or Methods).

Conventional English-language usage would have us employ the common world "need" as a verb (or in a verb sense) .to identify means, methods, activities, and actions and/or resources we desire or intend to use.[6]

Terms such as "need to," "need for," "needing," and "needed" are common, conventional, and destructive to useful planning. What? [7]

We have already noted this as Critical Success Factor 2. In order to do reasonable and justifiable planning we have to

(1) *Focus on Ends and not Means*, and thus
(2) *Use "need" as a noun*. Need, for the sake of useful and successful planning is only used as a noun, as a gap between current and desired results.

If we use *need* as a noun, we will be able to not only justify useful objectives but we will also be able to justify what we do and deliver on the basis of costs-consequences analysis.

We will be able to justify everything we use, do, produce, and deliver. It is the only sensible way we can demonstrate value added.[8]

---

[6] Because most dictionaries provide common usage not necessarily correct usage, they note that "need" is used as a noun as well as a verb. This dual conventional usage doesn't mean that it is useful. Much of this book depends on a shift in paradigms about "need." The shift is to use it only as a noun . . . never as a verb or in a verb sense.

[7] As hard as it is to change our own behavior (and most of us who want others to change seem to resist it the most ourselves!) it is central to useful planning to distinguish between Ends and Means.

[8] Sloppy word usage is comfortable but deceptive. How can one justify a statement "we 'need' to do a needs assessment" when the only practical needs assessment is about gaps in results, not gaps in means or resources. Words have meaning.

Bibliography

Barker, J. A. (2001). The *New Business of Paradigms* (Classic ed.). St. Paul, MN: Star Thrower Distribution. Videocassette.

Bernardez, M. (2005). Achieving Business Success by Developing Clients and Community: Lessons from Leading Companies, Emerging Economies and a Nine Year Case Study. *Performance Improvement Quarterly, Vol. 18*, Number 3. Pp. 37-55.

Bernardez, M. (2006a) *Tecnología del Desempeño Humano.* Chicago, IL: ITSON Business Global Press.

Bernardez, M. (2006b) *Desempeño Organizacional.* Chicago, IL: ITSON Business Global Press.

Bernardez, M. (2008). *Capital Intelectual.* Bloomington, IN., AuthorHouse,

Brethower, D. (2006) *Performance Analysis: Knowing What to Do and How.* Amherst, MA., HRD Press.

Clark, R. E. & Estes, F. (2002). *Turning Research into Results: A Guide to Selecting the Right Performance Solutions.* Atlanta, GA. CEP Press.

Drucker, P. F. (1993). *Post-Capitalist Society.* New York: Harper Business.

Gilbert, T. F. (1978). *Human competence: Engineering worthy performance.* New York: McGraw-Hill.

Guerra, I (2003). Key Competencies Required of Performance Improvement Professionals. *Performance Improvement Quarterly.* 16 (1).

Guerra-Lopez, I. (2007). *Evaluating Impact: Evaluation and Continual Improvement fo Performance Improvement Practitioners.* Amherst, MA. HRD Press.

Kaufman, R. A. (1968). A system approach to education--derivation and definition. *AV Communication Review*, 16, 415-425.

Kaufman, R. A., Corrigan, R. E., & Johnson, D. W. (1969). Towards educational responsiveness to society's needs: A tentative utility model. *Journal of Socio-Economic Planning Sciences*, 3,151-157.

Kaufman, R. A. (1972). Educational *system planning*. Englewood Cliffs, NJ: Prentice-Hall. (Also Planificacion de systemas educativos [translation of Educational system planning]. Mexico City: Editorial Trillas, S.A., 1973).

Kaufman, R., & English, F. W. (1979). Needs *assessment: Concept and application*. Englewood Cliffs, NJ: Educational Technology Publications.

Kaufman, R. and Forbes, R. (2002). Does Your Organization Contribute to Society? *2002 Team and Organization Development Sourcebook*; McGraw-Hill NY, 213-224.

Kaufman, R. & Lick, D. (Winter 2000 – 2001). Change Creation and Change Management: Partners in Human Performance Improvement. *Performance in Practice*, 8-9.

Kaufman, R, & Unger, Z. (2003: Aug.) Evaluation Plus: Beyond Conventional Evaluation. *Performance Improvement*, Vol. 42, No. 7, Pp. 5-8.

Kaufman, R., Guerra, I., and Platt, W. A. (2006) *Practical Evaluation for Educators: Finding what Works and What Doesn't*. Thousand Oaks, CA: Corwin Press/Sage.

Kaufman, R. (1998).Strategic Thinking: A Guide to Identifying and Solving Problems. Revised. Washington, DC & Arlington, VA: The International Society for Performance Improvement and the American Society for Training & Development. Also, Spanish edition, El Pensamiento Estrategico. Centro De Estudios: Roman Areces, S.A., Madrid, Spain.

Kaufman, R. (2000). Mega Planning: Practical Tools for Organizational Success. Thousand Oaks, CA. Sage Publications.

Kaufman, R. (2006). *30 Seconds That Can Change Your Life: A Decision-Making Guide for Those Who Refuse to be Mediocre*. Amherst, MA. HRD Press Inc.

Kaufman, R. (2006). *Change, Choices, and Consequences: A Guide to Mega Thinking and Planning*. Amherst, MA. HRD Press Inc.

Kaufman, R. (2006). Seven Stupid Things People Do When They Attempt Strategic Thinking and Planning, in Silberman, M., & Phillips, P. *The 2006 ASTD Organization Development & Leadership Sourcebook. Alexandria, VA.*

Kaufman, R. (2006:Aug.) Failure. What it is and how to invite it. <u>Performance Improvement.</u>

Kaufman, R., Watkins, R., & Leigh, D. (2001). <u>Useful Educational Results: Defining, Prioritizing, and Accomplishing</u>. Lancaster, PA., Proactive Publishers.

Kaufman, R, Oakley-Browne, H., Watkins, R., & Leigh, D. (2003). *Practical Strategic Planning: Aligning People, Performance, and Payoffs*. San Francisco, Jossey-Bass/Pfeiffer.

Kaufman, R. & Bernardez, M. (2005) Eds. *Performance Improvement Quarterly*, Special invited issue on Mega planning. Volume 18, Number 3. Pp. 3-5. http://www.ispi.org/publications/piqtocs/piq18_3.htm

Kaufman, R, & Guerra-Lopez, I. (2008) *The Assessment Book: Applied Strategic Thinking and Performance Improvement Through Self-assessments*. Amherst, MA. HRD Press Inc.

Lick, D. & Kaufman, R. (2000). Change Creation: The Rest of the Planning Story. In *Technology-Driven Planning: Principles to Practice*. J. Boettcher, M. Doyle, & R. Jensen (Eds.). Ann Arbor, MI. Society for College and University Planning.

Mager, R. F. (1997). *Preparing Instructional Objectives: A Critical Tool in the Development of Effective Instruction*. (3rd ed.). Atlanta: Center for Effective Performance.

Peters, T. (1997). *The Circle of Innovation: You Can't Shrink Your Way to Greatness*. New York: Alfred A. Knopf.

Prahalad C. K. (2005). *The Fortune at the Bottom of the Pyramid: Eradicating Poverty Through Profits*. Upper Saddle River, NJ. Wharton School Publishing/Pearson Education, Inc.

Rummler, G. A. (2004). *Serious Performance Consulting: According to Rummler*. Silver Spring, MD, International Society for

Performance Improvement and the American Society for Training and Development.

Senge, P. M. (1990). *The Fifth Discipline: The Art & Practice of the Learning Organization.* New York: Doubleday-Currency.

Watkins, R. (2007) *Performance by Design: the Systematic Selection, Design, and Development of Performance Technologies that Produce Useful Results.* Amherst, MA. HRD Press.

Roger Kaufman, PhD., CPT

Roger Kaufman is Professor Emeritus, Educational Psychology and Learning Systems, Florida State University. He is also Distinguished Research Professor, Sonora Institute of Technology (Mexico).

He is past president as well as honorary member for life of the International Society for Performance Improvement (ISPI) as well as the recipient of the Thomas Gilbert Award, ISPI.

He has recently been recognized by ASTD for Distinguished Contribution to Workplace Learning and Development.

He consults world-wide and is the author of 39 books and over 260 articles on strategic planning, needs assessment, and evaluation

Using technology and innovation for the planning social and economic transformation in a region of Mexico

By

Gonzalo Rodriguez Villanueva & Roberto Lagarda Silva

1. Antecedents

Humanity faces multiple challenges nowadays, including economic, political, cultural, and environmental. However, technology and social organization improvements can support regional innovation, which opens the opportunity to move towards prosperity and a society full of opportunities (Rodriguez, 2007).

But, how do we create a regional development plan based in innovative ecosystems allowing them to increase enterprise and well being competency?

Figure No. 1. Sonora South

Faced with new globalization challenges in Mexico, and in Sonora, the Sonoran Institute of Technology (ITSON) decided to redefine its social role in the community (Rodríguez & Guerra-López, 2005); it elaborated a new vision and a new mission and began a process allowing the reconfiguration of the relationship between the social and productive sector and ITSON. This process was formed mainly by identifying strategic sectors and defining high impact projects along with the government, enterprise and society.

With the support of the municipal presidents where ITSON is located, as well as the Quiriego municipality, along with economy and education secretaries, all of who are coordinated by ITSON, this plan seeks to support innovation through four ecosystems where regional competitive advantages can be developed, such as: biotechnology and agribusiness, software and logistics, ecotourism and sustainable development, and education and health.

Development of these innovative ecosystems requires the support of the university, which generates valuable knowledge, but flexible policies are required from the government, allowing the creation of solutions to the main problems in the communities; the support is also required from the investment of companies, which will generate opportunities for innovation and technological development that will help increase productivity, product and service quality.

2. Intellectual capital, competitiveness and economic development

Innovation has been very important for economic growth, and higher education are critical actors for that accomplishment.

Between 1970 and 1995, more than half of the total production growth in developed countries where due to innovation.

In these same countries, Higher Education Institutions (HEI) received a great amount of the investment, according to the Organisation for Economic Co-operation and Development (OCDE, 2007); innovation has been a key catalyst in competitiveness and economic growth.

The real competitiveness according to (López-Claros, Porter, Sala-i-Martin, & Schwab, 2006) is measured by productivity.

Productivity supports high salaries, and a return on investment and a high quality of life. Productivity is the goal, not exports.

Social and Organizational Performance Review

The global competitiveness index (GCI), developed by the World Economic Forum (WEF), and establishes 12 pillars that must be adhered to by countries and companies to develop their competitiveness in world economy through three phases:

1. Countries and companies focused on basic factors of production
2. Countries and companies focused on efficiency
3. Countries and companies focused on innovation

The three phases must be completed in the order shown above. Innovation can never be developed while there are no base conditions (phases 1 and 2), this means institutions with intellectual property, skilled workers and entrepreneurs, their economies with efficient markets and proper infrastructure.

Countries and companies must respect this process to achieve a greater competitiveness in the future.

Figure 2. The 12 pillars of competitiveness

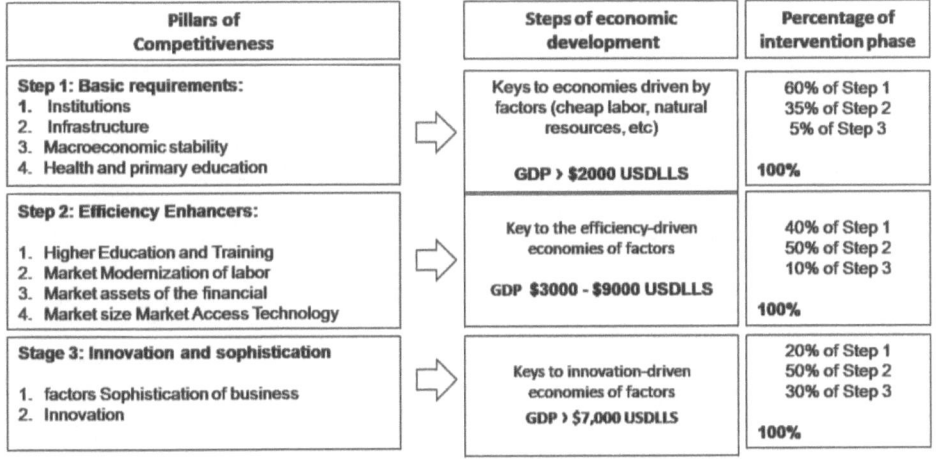

Pillars of Competitiveness	Steps of economic development	Percentage of intervention phase
Step 1: Basic requirements: 1. Institutions 2. Infrastructure 3. Macroeconomic stability 4. Health and primary education	Keys to economies driven by factors (cheap labor, natural resources, etc) **GDP > $2000 USDLLS**	60% of Step 1 35% of Step 2 5% of Step 3 100%
Step 2: Efficiency Enhancers: 1. Higher Education and Training 2. Market Modernization of labor 3. Market assets of the financial 4. Market size Market Access Technology	Key to the efficiency-driven economies of factors GDP $3000 - $9000 USDLLS	40% of Step 1 50% of Step 2 10% of Step 3 100%
Stage 3: Innovation and sophistication 1. factors Sophistication of business 2. Innovation	Keys to innovation-driven economies of factors **GDP > $7,000 USDLLS**	20% of Step 1 50% of Step 2 30% of Step 3 100%

The main characteristic of the first phase is having 60% of the PIB invested in basic requirements, generating an annual income less than

2.000 dollars; the second phase uses 50% of the PIB investment in efficiency, generating an annual income between 3 and 9 thousand dollars; and the third phase uses 30% of the PIB investment in innovation and development, which generates high added value products with an annual income greater than 17 thousand dollars (figure 2).

According to Sirkin, Hemerling and Bhattacharya from the Boston Consulting Group, there is a new name for globalization, and it is the global concept, which states that if we look for support in the proper technology use and we have a proper internationalization strategy, we all can compete against anyone, from anywhere and in anything. It is no longer a requirement to be in a developed country to generate high added value products and services.

If we consider the roles innovation and the global idea play, we are able to identify certain elements that have allowed some regions of the world to accelerate their growth, which is related in every case of technological cities (or the *technopolis*, translating to the Greek root).

Technopolises, or innovative ecosystems, are planned developments, supported by visionary actors along the private sector as well as the public sector, which their purpose has been to guide and control through innovation, and local and regional development processes that improve society's economic performance.

So, if we want to generate a plan for regional development based on innovation, new requirements appear for companies and universities seeking to create high added value products and services, as well as intellectual capital to satisfy the new needs of society.

Traditional organization models focused on the production and distribution of great products and services, and universities emphasized the graduation of large numbers of professionals trained to standard curricula. Such system turned out to be rigid, costly and slow to respond to new societal demands.

The *intellectual capital* concept according to Bernardez (2007), has *active components*: competencies, knowledge and rational and emotional aptitudes in a technology-scientific and artistic sense, required to create new ideas, concepts, products and systems – the so called intellectual assets – and a passive components: generated products that transform into cultural patrimonies and eventually into intellectual property.

There are four key factors that explain the intellectual property process:

1. Freedom of thinking, market and commerce
2. Expansion of superior education and investment in human capital
3. Organization of social and market ecosystems for the application and diffusion of thinking
4. The role of the superior education system in the context of society

3. The role of the superior education system in the context of society

The transition from one stage of economic and social development to another is explained by the gradual advance of knowledge and application of new technologies to produce goods and services that increase the quality of people's lives.

Pink (2006) explains the value shift from an *agrarian society* that prioritizes the production of tangible commodities to the *conceptual society* which sees people as the main factor of production, as creators of intangible value and meaning.

Sharmer and Käufer (2000), report that changes in higher education have been made through three phases: from the *scholastic* to the *classic* university, and from the *classic* to *postmodern*.

The *postmodern university* works within a conceptual society as a practice, research, and teaching unit; Table 1 shows both proposals divided by a kind of society.

Table 1. The phases of the evolution of the economic system and of the university by kind of society

Kind of society	Economic system Critical success factors (priority) Pink, D. (2006)	Higher Education System Sharmer & Käufer (2000)
Agrarian Society	1. Raw material 2. Agriculture	*University medieval scholastic (Teaching Unit)* **Education:** A lesson from the teacher. The student learns to listen and think.
Industrial Society.	1. Machines and tools 2. Raw material 3. Factory Workers	*Humboldt University Classic (Teaching and Research Unit)* **Education:** As above, adding studies seminar. The student also spoke. **Research:** Researchers working in solitude and freedom."
Information Society	1. Information and communication 2. Machines and tools 3. Raw material 4. Knowledge workers	*Universidad del Siglo XXI (Unit of practice, research and teaching)* **Education:** As above, plus infrastructure to: initiate, develop and undertake projects.
Conceptual Society	1. People create meaning. 2. Information and communication 3. Machines and tools 4. Raw material	**Research:** Research in action. Research consortia, clinical research and community action. **Implementation:** Co-creation in strategic companies, trusts, venture capitalists, incubators.

The Table 1 shows how to prioritize at the economic level and at the educational level; we can see that in the *agrarian society*, raw material and agriculture were the most important things from the economic perspective.

In the agrarian society, the teaching process determined what the student should learn, how to listen and how to think.

In the industrial society, the priority was the use of machinery and tools from the economic development perspective; the favored kind of teaching is specialized and technical, and research is individual.

In the information and communication society, technologies play an important role in the economic development, there are more resources in schools to create projects, and research is community based.

Finally, the *conceptual society* prioritizes people creating meaning in an economic system, research is community and project based, it is important to foster the strategic co-creation of companies, company incubators, and seeks to balance investment returns for the economy and society.

The challenge of universities is to adapt to new world context; for Prahalad and Ramaswamy (2005), the convergence of deregulation, emerging markets, technology and industries convergence has changed many facets of the business world.

These factors have also changed the nature of companies. Firms can fragment their value chains in a way that could not be done before. The physical and non-physical parts of a corporation can be divided.

Earlier tendencies allowed the business world a new way of creating value: one in which value is not created in the firm and later interchanged with the client, but rather co-created by the firm and the client. (Prahalad & Ramaswamy, 2005)

To achieve this, we must to work under a new university model including, besides its traditional teaching and research missions, a third mission: *the commitment to the communities they support.*

UNESCO lists *eight global characteristics* universities must contemplate as education, research, innovation, and technological development spaces:

1. See their mission transcending beyond the State-Nation frontiers, focusing their education service in a global perspective.

2. They have more intensive researches pursuing the scientific model.

3. Professors, as new knowledge producers, assume new roles, going beyond traditional paths to become interdisciplinary and international academic members guiding their research towards real world problems.

4. Research organizations are costly, therefore, universities go beyond the government's support and students cost to diversify their sources of income.

5. New relationships are being created amongst universities, governments and corporations to move forward the economic development and knowledge production to favor well being.

6. Universities are adopting recruitment strategies around the world for students, professors and academic personnel.

7. Institutions require greater internal sophistication towards research, such as interdisciplinary centers, integration of research elements in training programs, as well as a greater technological infrastructure.

8. Universities work with non-government organizations and multi-government organizations to support collaborative research, students and professors, as well as the validation of their international status.

According to the Kellogg Commision (2001) institutions committed to their local communities and regions must redesign their teaching, research, extension and services functions to obtain greater participation within their communities, by accomplishing three objectives:

1. *Be organized to respond to today and tomorrow's needs* of the student, not yesterday's.

2. *Enrich the student's experience by incorporating research and social commitment and offer practice opportunities* for the students so they can be prepared for the real world

3. *Put to work their critical resources (knowledge and experience)* in problems facing the communities they support

It is well known that superior education in Mexico has evolved, reforms in universities have not been enough to create an educational model that responds to today's needs.

So the *National Association of Universities and Institutions of Higher Education (ANUIES),* contemplates education must be understood as a *public good,* so that IHEs act with responsibility, quality and efficiency in their academic and social tasks, the country will be in better conditions towards a more competitive economy, a more fair and balanced society, and a more democratic political system.

4. Innovative ecosystems as creators of regional development

A region can be composed of a social and market ecosystem for the application and diffusion of new technology and ideas because they can put together different interests of the political sector at the local level in science and technology, industry performance, education and skills, health, and culture.

Therefore, the concept of regional economy based on innovation can be effective with the instrumentation of proper public policies.

Today, the old development model in which the government promotes economic growth through decisions and incentives in public policy, is no longer operational.

The new economic development model for society implies a collaborative process including the government (in its three levels), companies, universities and their research centers and social institutions through innovative ecosystems.

Defined broadly, *ecosystems* are structures allowing social and private sector actors, sometimes with different traditions and motivations and with different influence sizes and areas, to act jointly and create richness in a relationship.

Such ecosystems include a wide variety of institutions that coexist and complement each other. (Prahalad C.K., 2005, p. 65).

The *power of ecosystems* derives in large measure from their ability to attract, concentrate, train and support talent.

Modern urbanization expert Richard Florida (Florida,2009), describes how an ecosystem operates: *"well-educated professionals and creative workers who live together in dense ecosystems, interacting directly, generated ideas and turn them into products and services faster than talented people from other places can"* (p.6).

Ecosystems are spaces with specific vocations for social and economic development where clusters are located.

According to Porter, the model of cluster defines clusters as a group of organizations geographically close and interconnected by common and complemented elements.

Clusters are considered strategic initiatives for regional development based on innovation (Anderson, Schwaag, Sorvik, & Wise, 2004)

The benefits of clusters according to Hernandez, Fontrodona, & Pezzi (2007), are the following:

- *In productivity*: better access to production factors and skilled personnel; they increase knowledge of productive processes; facilitate the presence of specialized suppliers and favor access to technology.

- *In innovation*: the coexistence of competency and cooperation stimulate innovation; greater diffusion of technology; the presence of different companies in the value chain favors clients' needs.

- *In company creation:* favors the perception of business opportunities and therefore, stimulates spin-offs; supplier preservation, skilled personnel, support institutions, reduced input barriers, ease of information transfer, increased entrepreneurial dynamism and business opportunities.

For Etzkowitz (1997), universities, the productive sector and the government, have shown during the past years, they can create synergies to support regional development and this is done in development phases as shown in Table 2.

Table 2. Conceptual framework for economic development based on knowledge (Etzkowitz, 2002).

Steps of development	Characteristics
Creating an area of knowledge	Will focus on "regional innovation environments" where different actors are working to improve local conditions for innovation by focusing on R & D and other relevant transactions.

Creating a consensus	Ideas and strategies are generated in a 'triple helix' of relationships between multiple institutional sectors (academic, public, private).
Creating an innovative	Attempt to articulate the goals of the previous phase, establishing and / or attract venture capital and private-public (a combination of capital, expertise and knowledge of business) is a priority.

This represents a different way that relationships have existed between public, private and academic sectors, which traditionally have worked independently with each other (Etzkowitz, 2002).

The world competence also exerts an influence between IHE[9]s and their region. Globalization and TICs forces contribute to eliminate distances (Friedman, 2005).

The role of IHEs must promote teaching and research, and provide infrastructure to attract and retain the best researchers and professors of universities.

At the same time, regions must attract external investments based on knowledge, support local companies which desire to work in an international level, attract and retain creative talent.

There are multiple examples that show the important role of universities and research centers for the creation of innovative ecosystems and for the support of regional development:

- *The Massachusetts Institute of Technology (MIT)* uses organizations as laboratories for scientific and technological applications.

[9] Institutions for Higher Education

- *The University of Stanford,* located in Palo Alto, California, developed information technologies
- *Cambridge University, in the United Kingdom* worked in biotechnology

- *The Triangle Research park in North Carolina* has worked in biotechnology, computers and electronics

- *The University of Twente, in Holand* concentrates computers, embedded systems and information technology; and

- *The Technological Institute of India* has business centers for clusters located in special zones, such as: Bangalore, Bombay, among others.

In Mexico, there is movement towards a greater IHE participation in the development of new knowledge and in the social and economic development.

This shows in the *Programa Sectorial de Educacion*[10] *2007-2012,* which seeks a better relevance of the IHEs' educational programs.

Likewise, the operational strategic model from National Council for Science and Technology (CONACYT), proposes a better organized work for science, technology, and innovation development between IHEs/Research Centers, entrepreneur sector and government, related to the Triple Helix model, as previously mentioned.

The new budget scheme CONACYT (2007) supports, shown in Figure 3, emphasizes innovation and technology transfer by research centers, through technological parks and innovative partnerships.

---

[10] Regional Educational Program

Figure 3. CONACYT's programs and sub-programs' organization

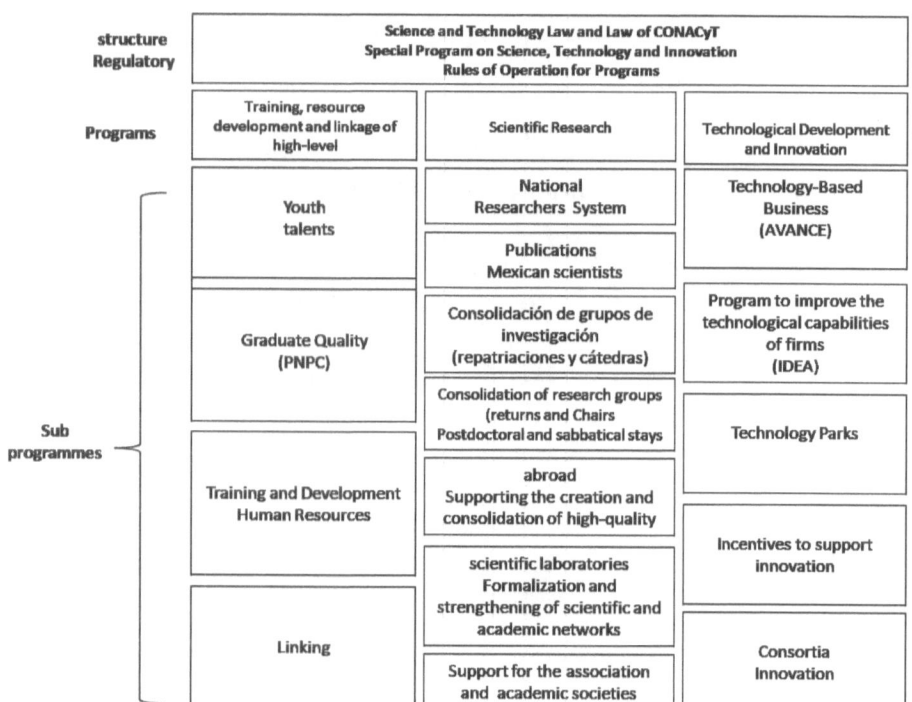

A key university role in developing innovation is creating and nurturing a "knowledge chain", from relations to the number of ideas in different phases

This implies the consideration of basic research, as well as technological development and knowledge application, and finally products and services development where companies also participate.

5. Creating an economic strategy to promote regional development based on innovative ecosystems.

The creators of public politic from the OrOCDE, state that their countries must be more competitive if they seek to maintain their economic position related to other economies and respond to challenges such as productivity gaps, competency for investment, the fast adoption of new technologies, and ecommerce, as well as the use of local resources.

If we seek to increase regional competitiveness, a new territorial paradigm is required with an integrating focus of the sector kind, the objectives seek to promote regional competitiveness, the tool is investment, the government stops being an actor and becomes multiple, that is, governments, non-governmental organizations and companies.

Table 3, shows the differences between the old and new territorial paradigm.

Table 3. Territorial paradigms

	Territorial paradigms	
	Old	**New**
Focus	Sector	Integrated
	Neutral value	Added value
Goals	Compensation	Promoting competitiveness
Tools	Financial subsidies	Investment
Actors	Central government	Multiple levels of government, NGOs and Corporations

The new territorial paradigm presented by OCDE's expert Nicola Crosta, states five key elements a strategic plan must contain for regional development based on innovation.

- *Government:* determination of priority areas that must orient long term public policy

- *University:* Strategic development through the creation of innovative ecosystems that generate high added value products and services

- *Cluster or value chain development*: provide, promote, and support companies allegiances, and the integration of value chains to facilitate the internationalization of regional products and services

- *Creation of technological infrastructure for innovation*: synergies within the region and alliances around the world

- *Education and training programs:* facilitating the incorporation of personnel into innovative ecosystems for regional development

Figure 4 shows relationships between different variables focused on territorial development, we can see this basically depends on five key variables:

(1) research and development,
(2) information technology infrastructure,
(3) incubation,
(4) human resources and
(5) Government.

These five variables are interrelated.

The infrastructure variable is related to the relationships given between electronic companies and their demand for infrastructure, but at the same time they seek private and public investment for their growth.

When they grow, they support regional development and *"techno polis"*- universities grow as well, seen as geographical clusters of knowledge, skills and strategic alliances giving the community a competitive advantage supported by research and technological development to increase human resources through training and education.

Figure 4: Causal relationships for regional development (Nicola Crosta)

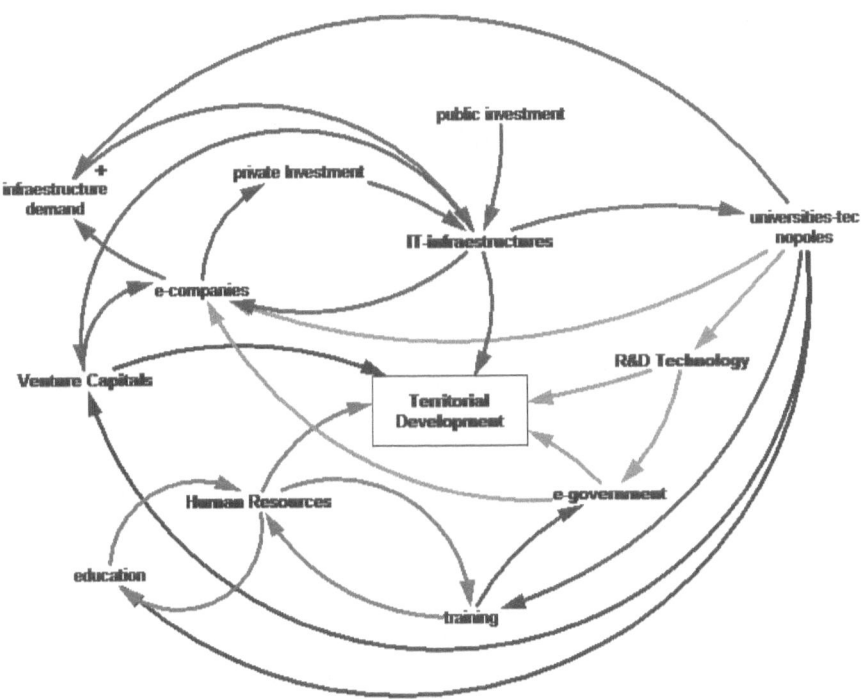

Kaufman (2000, 2006) proposes the organizational elements model (OEM), which defines links and aligns whatever the organization uses, produces, and delivers to external clients and added value to society.

Each element has an association to a planning level.

Strategic planning (and thinking) begins with the mega focus while tactic planning begins with macro and operational planning begins with micro.

Table 4 shows the three planning levels and their description.

Table 4: The three planning levels and their description (Kaufman, 2000)

Name of the organizational element	Name of the level of planning and focus	Brief description	Level of Planning
Outcomes	Mega	Results and their consequences for external clients and society.	Strategic
Outputs	Macro	The results an organization can or does deliver outside of itself.	Tactical
Products	Micro	The building- block results that are produced within the organization.	Operational
Process	Process	The ways, means, activities, procedures, and methods used internally.	_____
Inputs	Input	The human, physical, financial resources an organization can or does use.	_____

Mega results include those social indicators that impact in the vision through the growth and development of society, market and customers, stability and social progress among others. Macro results, are value and the income generated and measured in terms of its contribution to economic sustainability of the organization or its prestige and acceptance by its customers; Micro results are measured in terms of domestic products of the organization.

The micro products are transformed in macro results, when they are recognized by customers and sponsors to generate revenue (from sales or investment), recognition (such as certificates for enabling the institution). The Macro results are changed in Mega results when they impact in social indicators associated to the Vision.

Bernardez (2007) proposed a methodology budgeting Mega, which is considered not only the result of micro projects and their costs, but the results Macro-institutional revenue and results-Mega-income social structure of the central budget is a Mega business case that defines four major areas along a period of 3 to 5 years.

6. Methodology proposal to generate the integral development plan for Southern Sonora

The state and municipal governments, companies and ITSON have committed to generate an integral development plan for the Southern region of Sonora, in which it is possible to accelerate the economic and social development of a region from the creation of innovative ecosystems allowing them to convert their comparative advantages into competitive advantages through the regulatory framework of their institutions, process efficiency, and the support for cluster creation and a proper environment for innovation. Figure 5 shows the systemic model to support the regional development based on innovation.

The model includes the following elements:

- Value proposal for the region: vision statement and strategies allowing the region to be more competitive
- Innovative ecosystems: includes four fundamental elements, related to: institutional and social framework; operational efficiency of the products and factors market; business sophistication degree; innovation support.
- Potential for economic and social development: related to the region's scientific-technological capacity, as well as the

cultural patrimony and protection of intellectual and material property rights

- Strategic initiatives: capacities and opportunities (strategic areas) in the region, to convert comparative advantages into competitive advantages
- Innovation and development centers: organizational structure formed by social and academic advisors, for the development of strategic projects generating interdisciplinary solutions with high added value to support the creation of strategic initiatives and innovative ecosystems

***Figure 5.* Systemic model to support regional development based on innovation**

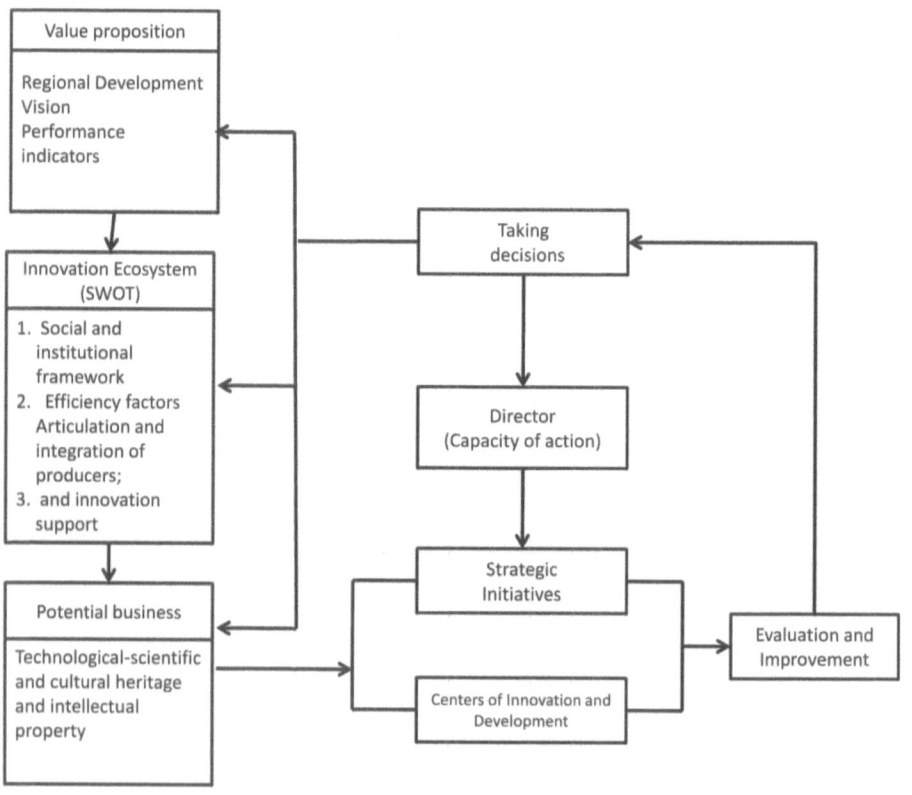

The methodological process contemplates the six following phases shown in figure 6, beginning with a shared vision; needs identification; sensibility and conceptualization phase; environments creation; and development of SWOT and brainstorming; work plan elaboration; objectives, strategies, and goals definition; and finally the implementing phase; elaborating projects, performance implementation, measurement, and evaluation considered as feedback for the methodology and continuous improvement which must be consistent with the established vision.

Figure 6 shows the complete process.

Figure 6. **Proposed methodology for the regional development plan**

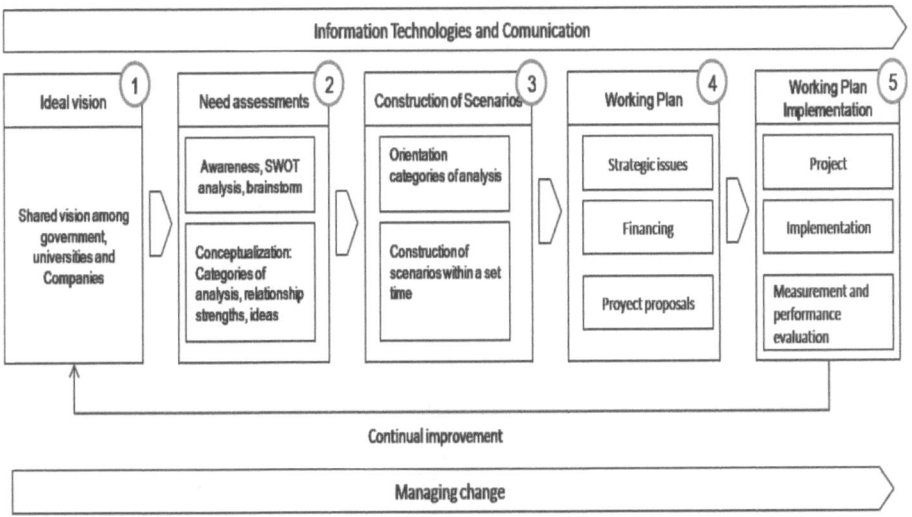

Shared vision:

We defined a shared vision derived from an ideal vision.

Main allies such as government, companies, and universities use the shared vision as a common "guiding star" guiding and aligning their long-term efforts, across multiple government periods.

7. Strategic planning methodology application results for regional development based on innovative ecosystems.

The participation of the allies in this process has enabled the following results:

Shared vision:

Municipal development plans 2006-2009 from Guaymas, Cajeme and Navojoa were analyzed, as well as the Plan Estatal de Desarrollo 2003-009.

According to Rodriguez (personal communication, November 17, 2008) the steps were taken to strengthen the human capital system of the state and develop international networks and collaboration projects for the growth of the productive sector through innovation and knowledge development infrastructure.

Figure 7. Competitive agenda of the State of Sonora

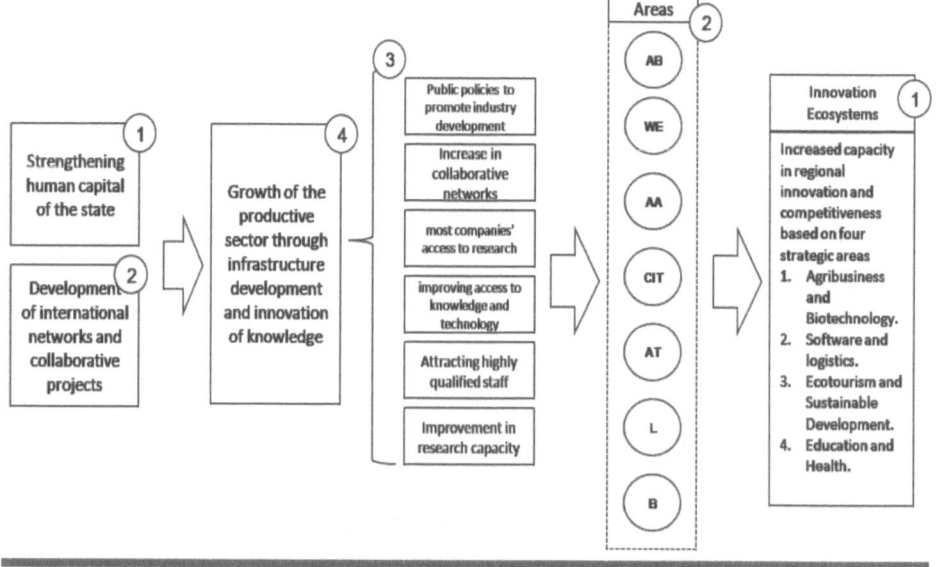

We look to support seven sectors: *Agribusiness* (AB), *Water and Energy* (WE), *Automobile and Aerospace* (AA), *Communication and Information Technologies* (CITs), *Alternative Tourism* (AT), *Logistics* (L), and *Biotechnology* (B), to increase regional capacity of innovation and competitiveness.

Likewise, the economy index was considered (IEC) for the state of Sonora (*Fundación este país*, 2008) where Sonora is ranked in 7th place among the nation.

Figure 8. Knowledge Economy Index for the State of Sonora (Fundación este país, 2008)

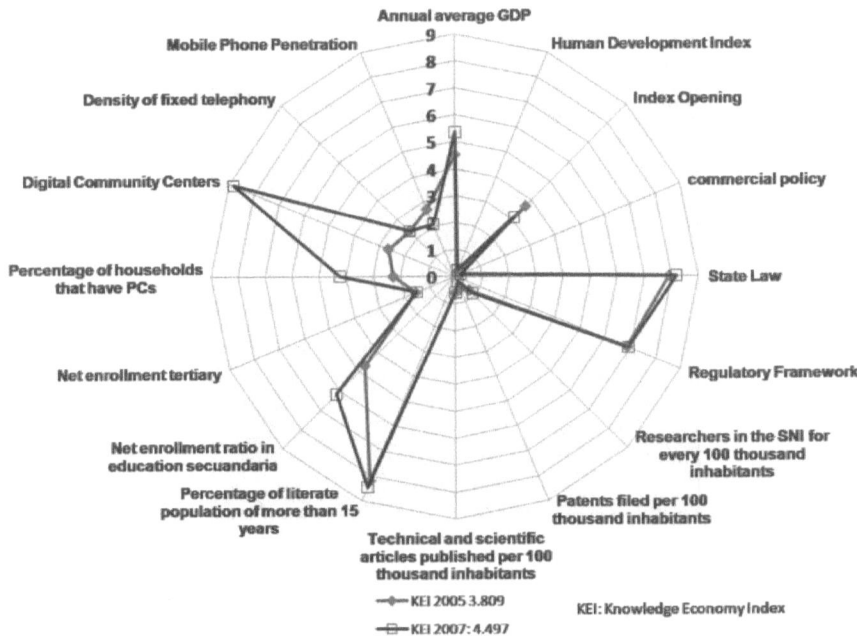

Considering the previous antecedents, the integral development plan for the Southern region of Sonora is the result of the commitment made by the three municipal presidents of Guaymas, Empalme, Cajeme and Navojoa, and Quiriego representing rural zones, in conjunction with Secretary for Education and Culture, the Secretary of State's Economy, and coordinated by the Sonoran Institute of Technology, during the second semester of 2008.

To ensure economic growth guides towards the regional development of the community, it is necessary to design and implement a public policy and a strategy to resolve poverty and inequality issues.

To accelerate the integration of the region into globalization and achieve continuous improvement of productive processes, an internationalization strategy is required to support the creation and implementation of organizations and institutions in society based on innovation and the creation of high added value products and services.

In relation to what was mentioned above, we define the following vision:

The southern region of Sonora is part of an economy and society allowing the state to have high economic dynamism enhancing quality of life by generating high added value products and services in a safe and healthy environment for talent and investment development. In a long term plan, we consider regional specialization for the generation of high added value products and services, as well as the creation of an information and communication technological infrastructure, and quality of life.

The long term plan is shown in figure 9.

Figure 9. Geographical localization of the southern region of Sonora

Sonora State

Map of Mexico

The state of Sonora, is the second largest in the country, its territorial extension is 184, 934 km ².

The southern region consists of Guaymas, Empalme, Bacum, Cajeme, Yecora, Rosario, Quiriego, Poplar, Navojoa, Benito Juárez, Etchojoa, Huatabampo.

South Zone

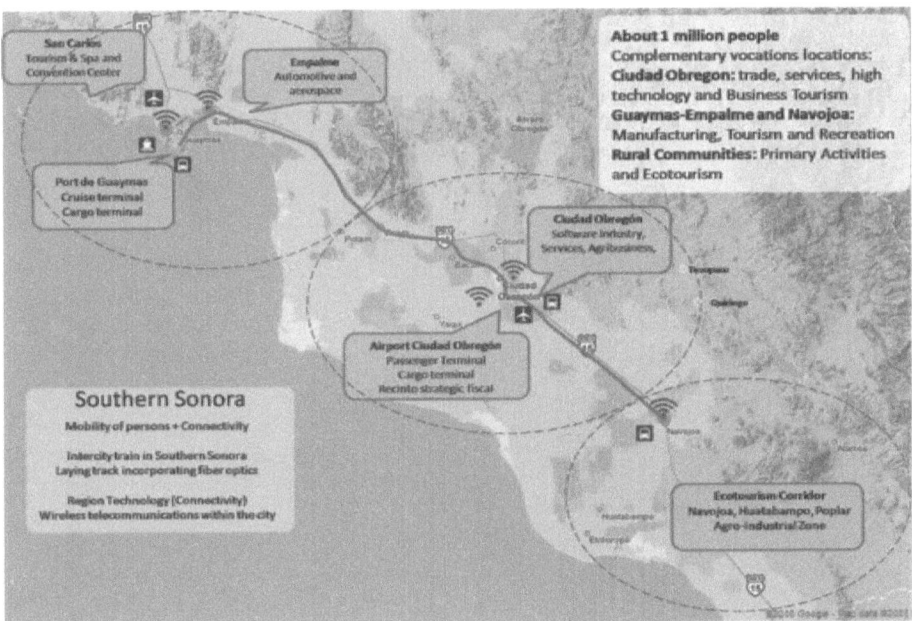

Sensitivity phase: A panel was developed to reflect and analyze the southern region of Sonora; the objective was to generate a

Social and Organizational Performance Review

participating space to reflect and analyze strengths, opportunities, weaknesses, and threats (SWOT) of the region.

Table 5 shows the results.

Table 5: SWOT analysis of the Integral Development Plan for the southern region of Sonora.

Strengths	Opportunities
Economic engines: • **Guaymas port.** • **Maquiladora industry.** • **Eco-touristic attractions.** • **Main export production by its highest sanity level (pork, shrimp, vegetables)** • **Notable region potential for development of fish production** • **Cluster machined aerospace specialist**	• Design a regional economic vision based on innovation and the use of TICs. • Create a high tech manufacturing industry to support the national and international markets • Incorporate rural populations in value activities according to their natural resources to support eco-tourism • Develop local supply • Develop agribusiness chain through the industrial and commercial transformation of farming and fish into high added value products.
• **Privileged strategic location: access to the United States (Arizona and California), Pacific river basin (emerging Asian markets) and the Canamex corridor.**	• Consolidate airport projects (Cd. Obregon) and support the proposal of a fast train facilitating the mobility of economic resources between Guaymas and Navojoa.
• **Qualified workforce and proper work environment for investment.**	• Generate jobs that use knowledge and education provided by companies, universities and regional research centers.
• **Economic growth above the Mexican average.**	• Advantage of universities and regional research centers to generate safe and healthy communities that can attract productive investments.
• **Capacity to develop and to exert**	• To favor the production of alternative

proper technological solutions for water and energy production. • **Basic urban infrastructure.** • **Support from ITSON (proposals to develop legal initiatives towards tourism and science).**	water and energy sources (solar and wind power), to favor sustainable development. • ITSON as a promoter of different sectors and institutions.
Weaknesses	**Threats**
• **Population decrease from 1990 to 2000, especially in the rural sector.** • **Reduced public services** • **Negative circumstances due to natural phenomena**	• Breaking the international order • Global economic crisis (economic flux reduction, jobs and salaries decrease, main products price decrease).
• **Water deficit in Guaymas** • **Reduced productive diversification and low farming modernization index** • **Low university/company links.**	• Polluting chemical use in main activities.
• **Lack of modernization and maintenance in communication.** • **Military checkpoints.**	• China and U.S. competence, and Sinaloa, Baja California and Chihuahua
• **Consumerism, obesity, inadequate use of personal time and low life perspective.**	• Insecurity increase.

Conceptualization phase: essential categories are defined derived from the vision: territorial integration (cluster creation and commercial intelligence and regional strategic planning); logistics (local suppliers development); infrastructure for social development (increase financial capacity, services expansion) and; government model (evaluate the social performance improvement).

Social and Organizational Performance Review

Orientation phase: based on the vision's essential categories, an instrument was generated to identify the sustainable development degree of organizations, shown in table 6.

Table 6: Intervention priorities for the regional integration plan.

Statement	Short term	Half term
1. **Identified on the basis of regional comparative advantages (strengths) areas of opportunity for the formation of clusters** (quality of life).	16	3
2. **Designing technological packages that allow the generation of products and services with high added value** (innovation).	6	9
3. **Building the organizational processes that meet international quality standards** (best practices).	1	13
4. **Develop a monitoring program to support individual and organizational transformation** (talent).	8	2
5. **Develop local suppliers and marketers that are reliable** (service orientation).	5	8
6. **Outsource non-core activities: flexible structure and differentiated** (adaptability).	1	1
7. **Raising the financial, credit and insurance organizations of** (self-sufficiency).	1	8
8. **Share business intelligence and participate in strategic planning regional** (cooperation).	10	2
9. **Geographic expansion and broadening of the coverage of goods and services from brand positioning** (identity).	1	2
10. **Evaluate and improve organizational performance and social** (responsibility).	3	4
Total votes (members)	**52**	**52**

On the other hand, two responses were given for the analysis of the southern region of Sonora that was focused in these five areas: Guaymas, Cajeme and Navojoa, as well as the Education and Economic Sectors.

The first question: What elements and institutions must intervene for the development of value chains within the integral development regional strategic plan for the southern region of Sonora?

Results are shown in table 7.

Table 7: Value chains results.

Municipalities			Sectors	
Guaymas	Navojoa	Cajeme	Education	Economy
• **Tourism** • **Reactivation of railroad activity** • **Reorientation of maquiladora industry** • **Increase of fish and farming activity** • **Use land that has not been planted because of saline intrusion to plant wheat and produce ethanol and watering with wastewater treated Guaymas-Empalme.**	Agribusiness (wheat and oils) • Cattle • Miners • Software development • Logistics • Tourism.	• Agribusiness main products: fish, without affecting natural resources • Tourism • Ecotourism: gastronomy, crafts, medic and business. • Aerospace • Information and Communication Technology TICs Innovation	• Innovation • Education/jobs • Evaluation • Tourist and eco-tourism services • Technology and software • Fish development • Agribusiness • Educational services	• Agribusiness • Tourism • Manufacture

The second question is related to the main results that must be accomplished in a short and long term with an integral development plan for the southern region of Sonora; in short: increase quality of life of the region, generate more jobs, human development, cooperation between governments, improvement of social-economic indexes, Gross Domestic Product increase, better public security.

Social and Organizational Performance Review

Planning phase: four innovative ecosystems were defined: Biotechnology and Agribusiness; Software and Logistics; Eco-tourism and Sustainable Development; Education and Health. Table 8 shows the impact divided by ecosystem.

Table 8: Innovative ecosystems and their impact

Innovative Ecosystems	Impact
Biotechnology and Agribusiness	Clusters creation enabling the creation of high added value products and services (self sufficiency).
Software and Logistics.	Move from industrial economy towards knowledge economy (survival).
Eco-tourism and Sustainable Development	Migrate towards a rural-urban space to favor regional specialization and proximity to the workplace (health).
Education and Health.	Create community networks for science, technology, art, sports, and recreation to increase education quality (well being).

On the other hand, the relationship between knowledge areas and development phases was established, where the main focus is to achieve the productive sector integration in the knowledge chain (table 9).

Table 9 shows the existent relationship between defined innovative ecosystems in the strategic planning exercise for regional development and the strategic initiatives which ITSON has designed to support the generation of high added value products and services through innovation and development projects.

Table 9: Knowledge chain to integrate the productive sector

Area		phases			
	Capacity Scientific	Applied research.	New products and services	Innovation and development projects. Natural	
Natural Resources	Biotechnology and Agribusiness	• Experimentation and Technology Transfer Center • Research and Innovation in Biotechnology, Agriculture and Environmental • International District Agribusiness Small and Medium Enterprises	• Development of agricultural products. • Health insurance and innocuousness. • Supporting the commercialization of micro and small enterprises	• Develop high added value products and services through the integration of productive chains from the primary sector • Develop biotechnology products for the conservation of natural resources • Water and energy technology development •	
Engineering and Technology	Software and Logistic	• Technology Center for Integration and Business Development. • Software Factory Software • Technology Park.	• Support for micro and medium companies. • Technological solutions development • Support for technological companies.	• Logistic Operations Center development. • Design technological packages to increase competitiven	

				• ess in companies • Development of technological solutions based on alternative sources to achieve energy sustainability. • Development of technological solutions to improve educational performance.
Social and Administrative Sciences	Ecotourism and sustainable development	• Center for Research and Development of Water and Energy. • Business Incubator. • Ecotourism Corridor in southern Sonora.	• Eco-touristic corridors for the southern region of Sonora. • Identification of regional comparative advantages • Generation of new companies • Articulation and integration of communities.	• Economic and cultural development diagnostics from other regions in the world to increase the scope of products and regional services • Development of local value chains ensuring quality products and services • Development of Alternative

				Tourism Corridors for the southern region of Sonora • Development of intervention models for the community.
Education and Humanities	Education and Health	• University Center for Community Development. • University life. • Park Articulation and Transfer of Educational Technology	• Support to rural communities • Support and following of student trajectory • Universal access to digital technology	• Develop an articulation and integration enterprise plan to create productive chains in the primary sector of the Guaymas-Empalme region • Development of technological solutions to improve companies' competitiveness • Development of art and touristic corridors in: Bahia Kino, La Manga and Tobari and Centro Historico

If the strategic plan for regional development based on innovation is implemented, it will have a prospective impact on jobs and the economy in the year 2012, we will see an annual investment increase of 38.7% which would result in a job increase of more than 10,000 well paid jobs between 2008 and 2012.

Table 10 shows the total budget structure divided by innovation ecosystem.

Table 10: Business case by 2012 by Innovation Ecosystem

Innovative Ecosystem	Direct and indirect jobs			Investment (US Dollars)		
	2004	2008	2012	2004	2008	2012
1. Agribusiness and Biotechnology City	35	234	3,500	$250,000	$7,750,000	$25,214,286
2. Software and Logistics	85	370	4,300	$650,000	$11,250,000	$35,357,143
3. Ecotourism and Sustainable Development	40	110	1,500	$280,000	$2,666,667	$15,357,143
4. Education and Health (Digital City)	30	270	2,000	$150,000	$8,750,000	$20,571,429
Total	190	984	11,300	$1,330,000	$30,416,667	$96,500,000

8. Conclusions

Regions have the opportunity to develop when commitment exists, the triple helix model determines that universities and their research centers allied with governments and their three levels, as well as companies, are important alliances in this transformation process towards greater social survival, health, self-sufficiency and well being.

The creation of development plans focused on innovative ecosystems and strategic initiatives contemplates five elements: creation of an innovative regional system; support of the creation of productive chains; prioritize information technology, communication and transport investment supporting territorial integration; creation of a safe and healthy environment attractive to investors; and develop a government model based on a state policy and managed by projects.

Applying this kind of model will accelerate the region's economic growth and social development.

Bibliography

Allen L, H., & Kramer, W. J. (2007). *The Next 4 Billion: Market Size and Business Strategy at the Base of the Pyramid.* California USA: World Resource Institute.

Anderson, T., Schwaag, S., Sorvik, J., & Wise, E. (2004). *The cluster policies whitebook.* Sweden: Holbergs i Malmo AB.

ANUIES. (2006). *Consolidación y Avance de la Educación Superior en México.* México D.F.: ANUIES.

Bernardez, M. (2007). *Desempeño Organizacional, mejora, creación de incubación de nuevas organizaciones.* Bloomington, IN: Authorhouse.

Cumbre Mundial de la Sociedad de la Información. (Recuperado el 12 de Mayo de 2004, de http://www.itu.int/wsis/docs/geneva/official/dop-es.html).

CONACyT. ((2007). *Modelo Estratégico,* Recuperado de http://www.conacyt.mx/Acerca/Acerca_Introduccion.html).

Etzkowitz. (Recuperado el 5 de diciembre de 2008 de http://vlex.com/vid/triple-helice-socioeconomicos-regionales-

117510). *El modelo de triple hélice: una herramienta para el estudio de los sistemas socioeconómicos regionales europeos.*

Florida, R. (2009). How the crash will reshape America. *The Atlantic Monthly* , 23-36.

Friedman, T. L. (2005). *The World is Flat: A Brief History of the Twenty-firs Century.* New York, USA: Farrar Straus & Giroux.

Guerra, I. & Rodriquez, G. (2005). Educational Planning and Social Responsibility: Eleven Years of Mega Planning at the Sonora Institute of Technology (ITSON). Performance Improvement Quarterly, Vol. 18, Number 3. Pp. 56-64.

Hernández, J., Fontrodona, J., & Pezzi, A. (2007). *Mapa de los Sistemas productivos locales industriales en Cataluña.* Cataluña, España: OCDE.

Higher Education and Regions: Globally Competitive. (2007). Retrieved October, 06, 2007 http://globalhighered.wordpress.com/2007/10/06/higher-education-and-regions-globally-competitive-locally-engaged/).

Kaufman, R. (2000). *Mega Planning: Practical tools for Organizational Success.* USA: Sage Publications, Inc.

Kaufman, R. (2006). Change, Choices, and Consequences: A Guide to Mega Thinking and Planning. Amherst, MA. HRD Press Inc.

López-Claros, A., Porter, M. E., Sala-i-Martin, X., & Schwab, K. (2006). *The Global Competitiveness Report.* Geneva, Switzerlan: Palgrave macmillan.

Los objetivos del milenio. ((2000). Retrieve of http://www.un.org/spanish/aboutun/hrights.htm).

Pink, D. (2006). *Whole new mind: Moving from the information age to the conceptual age.* New York: Reverhead books.

Prahalad, C. K., & Ramaswamy, V. (2005). *The fortue at the bottom of the pyramid.* New Jersey, USA: Warthon School Publishing.

Rodríguez, G. (2007). *Contribuciones de las Instituciones de Educación Superior a la Generación de Consecuencias Sociales Positivas.* Bloomington Indiana: AuthorHouse.

Scharmer, C. O., & Käufer, K. (2000). *Universities as the Birthplace for the Enterpreneuring Human Being.* USA: Society for Organizational Learning.

Gonzalo Rodriguez Villanueva, M.S.

Gonzalo Rodriguez Villanueva has been the president of the Sonora Institute of Technology (ITSON) since 2003; he is a PhD candidate in economic sciences from the School of Economics, National Polytechnic Institute of Mexico. He has also participated internationally in the Mexican Universities' and European Universities Alliance (France, United Kingdom and Spain) –a forum dedicated to evaluation and perspective of common ground between Latin America higher education institutions and the European Union-; the North American Meeting for sustainable development in the US-Mexico border; at the fifth Ibero-American Summit of Presidents of Public Universities. Rodriguez Villanueva is member of the Mexican Council of Foreign Commerce, and president of the State Council for Social Participation, Transparence, Evaluation and Improvement Processes in educational institutions in Sonora. E-Mail: grodriguez@itson.mx

Ernesto Lagarda Leyva, M.S.

Ermesto A. Lagarda Leyva is the director of Institutional Planning Office at the Sonora Institute of Technology (ITSON); he has also participated as a federal evaluator of financial programs for Mexican universities and educational programs. As a member of the president's staff for support and planning, he is working on the 12 strategic initiatives, which involve participation of government, private sector and universities in regional development in the South of Sonora state.

He has also served as professor at Colombian universities and as a program manager with USAID and the ASU-ITSON Binational Consortium Partnership for desert Environmental Research at Arizona State University. He graduated from ITSON with a master's in industrial engineering (optimization of systems) and is currently in the last year of his doctoral program in strategic planning for performance improvement. His research interest include dynamic systems and performance for economic and social organizations. E-Mail: elagarda@itson.mx

Minding the business of business: tools and models to design and measure wealth creation

By

Mariano L. Bernardez, PhD., CPT

As thousands of concerned Americans learn each time a new housing, mortgage or stock market crisis strikes, wealth can be a fleeting proposition.

The continuous cycle of boom and bust that characterizes the dynamics of "creative destruction" that Joseph Schumpeter considered the hallmark and strength of entrepreneurial capitalism, makes it hard for investors, managers, consultants or employees using measurement tools based purely on conventional, backward-looking accounting or speculative stock performance indicators[11] to tell the *Googles* from the *Enrons*, the *Amazons* and *Grameens*[12] from *WorldComms, Fannie Maes and Freddie Macs, or* the *Newton Apples*[13] from the *iPods*[14] and *iPhones*..

Milton Friedman's famous truism about the nature and "social responsibility" of business being to increase its profits (Friedman M. , 1970)–popularly misquoted as "the business of business is business"-

---

[11] All the companies mentioned in this paragraph were stock market "stars" showing exceptional P&L and balance sheets. Half of them still are. The other half went bankrupt precipitating huge financial crisis and investors' losses

[12] Revolutionary bank created by 2006 Nobel Peace Prize Muhammad Yunus that lends under 20 dollar loans to 50 million clients in bottom-of-pyramid (BOP) markets (Yunus, 2003) (Prahalad C. .., 2005)

[13] Apple handheld device similar to the Palm PC that "flopped" in the market in 1984 and sent Apple stock spiraling down for many years. (Bernardez, Capital Intelectual: creacion de valor en la sociedad del conocimiento, 2008)

[14] Apple handheld products based on technology similar to the Apple Newton device that dominated the market in 2002-2006 and sent Apple stock to historic heights (Bernardez, Capital Intelectual: creacion de valor en la sociedad del conocimiento, 2008)

leaves us without a clue about how to create those profits, where and how they are made and thus, having a better chance to differentiate between a sound business proposition and more or less elaborated versions of "field of dreams" or plain scams.

What is then "the business of business"? How can planners and investors anticipate the true chances of failure or success of a business idea? How can we better filter scams?

"Conventional wisdom" –as John K. Galbraith labeled business thinking inertia- may quickly answer that the business of business is "making money" –whether revenue, cash flow or actual profits- and the way to measure a business ultimate performance and success are profit and costs projections presented as conventional, single-"bottom line"[15] business cases that monetize more or less wishful thinking based on generic assumptions about the future "business" usually presented as an extension of its past performance.

The problem with this approach –exemplified by the recurrence of Enrons, subprime mortgages and other scams window-dressed by these business cases' deceiving financial calculations- is that financial figures fail to explain *the business* in the "business case": how we create and deliver measurable and continuing value to the client, how much and how well our clients and their clients will do with our products and services. If a business case cannot explain convincingly these basic business concepts, a *"caveat emptor"*[16] sign should "flag" investors about the risks they are assuming.

But when Wall Street and Madison Avenue join forces to sell and hype their "business as usual" propositions, this seldom happens.

---

[15] "Bottom line" in financial lingo refers to the section of the profit and loss statement that reflect the net profit or loss before taxes. Following that model, business cases usually present a "bottom line" that reflects the yearly net result (profit or loss) , the "break even" point and calculates the Return on Investment –ROI- of the project. (Bernstein & Wilde, 2000)
[16] Latin for "Let the buyer beware". Under the doctrine of caveat emptor, the buyer could not recover from the seller for defects on the property that rendered the property unfit for ordinary purposes

Spaniards' cultural wisdom summarized in a sharp expression the actual value of such conventional, purely financial business cases: *"el papel lo soporta todo"* (*"everything looks good on paper"*)

A more contemporary and accurate approach to define a business proposition comes from factoring, analyzing and monetizing research-based data about how products and services actually increase clients, clients' clients and market's revenues, resources and wealth, by adding a second "top line"[17] that reflects clients and community revenue to the conventional income statement in order to determine where the money comes from and what kind of tangible benefits a client actually gets in exchange for what he/she paid.

This second top line measures not just profit extraction but value creation and renovation because it shows *how the company's products and services actually add value to clients, replace the natural resources utilized and strengthen the markets and communities that consume them* –something that would have set off the alarms at Enron when the company traders started to shut down Californian power utilities to gouge prices and sustain unrealistic return goals based on a purely financial business case made to gamble the stock market[18] -

The adoption of a double-bottom line business case methodology to plan and valuate business propositions is part of a 21st century business performance paradigm shift towards measuring wealth creation and maximizing social and environmental impact rather than just quarterly profit and short-term stock value[19].

---

[17] "Top line" in accounting and business lingo refers to the section of the profit and loss statement –usually at the top of the income statement- that describes total revenues and its sources. (Bernstein & Wilde, 2000) (Silbiger, 1993) (Bernardez, Desempeño organizacional, 2007)

[18] .Enron had created offshore entities, units which may be used for planning and avoidance of taxes, raising the profitability of a business. These entities made Enron look more profitable than it actually was, and created a dangerous spiral in which each quarter, corporate officers would have to perform more and more contorted financial deception to create the illusion of billions in profits while the company was actually losing money. (Ley Toffler, 2003) (Stiglitz, 2003)

[19] Quarterly profit and stock value fluctuations often create a reinforcing "loop" that drives companies towards self-destructive short-term management practices. (Bernardez, Rethinking the bottom line: Last frontier and first step for organizational change , 1996)

Social and Organizational Performance Review

This fast-growing consensus among business leaders and path-breaking companies springs from a series of forward-thinking pioneers such as Roger Kaufman (Kaufman, Corrigan, & Johnson, 1969), Peter Drucker (Drucker, 1985), C.K. Prahalad (Prahalad & Hamel, 1994) (Prahalad & Hammond, 2002), Michael Porter (Porter & Kramer, 2002) , McKinsey Group (Davis, 2005, May 26) and the research about wealth creation produced by the World Bank (International Bank for Reconstruction and Development / The World Bank, 1998)

This new approach to designing and planning business performance focused on "doing well by doing good" is rapidly gaining momentum among top, class-defining companies in the form of 10-year strategic initiatives, from General Electric "Green Initiative" to BP's "Beyond Petroleum" to Citibank's "Micro financing to the poor" or Microsoft and Intel Low cost PC for emerging countries.

Stock market valuation indexes provide further evidence that "doing good"[20] also seems to be the safest way for companies' stocks to do well in the short and midterm[21]: Figure 1 shows how Dow Jones Sustainability Index (DJSI)[22]-listed stocks of environmentally responsible, socially proactive companies have been outperforming consistently those of Dow Jones General Index (DJGI)[23]-listed, conventionally managed companies during the last 10 years.

---

[20] When we refer in this article and method to "doing good" we are not talking about companies engaging in conventional philanthropy, but to actively deliver products and services created to produce measurable financial, economic or social benefits for specific clients, their clients and their markets such as to help them to pay for our products and services in the long run. We are not talking philosophy or politics: we are talking plain microeconomics." Hence our motto: "What is good for GM's client is good for GM", *not the other way around"*.

[21] The time span of a business case is usually 5 to 10 years (Afuah, 2004) (Bernardez, Presupuesto Mega, 2007)

[22] Corporate Sustainability is a business approach that creates long-term shareholder value by embracing opportunities and managing risks deriving from economic, environmental and social developments. Corporate sustainability leaders achieve long-term shareholder value by gearing their strategies and management to harness the market's potential for sustainability products. The DSJI measures and compares the performance of companies pursuing a sustainable strategy with the average Dow Jones General Index (DJGI).More information at: http://www.sustainability-indexes.com/07_htmle/sustainability/corpsustainability.html

[23] Dow Jones General Index reflect the performance of conventionally-managed companies, based on purely financial indicators.

Figure 1: Dow Jones Sustainability Index outperforms Dow Jones General Index (2000-2005)

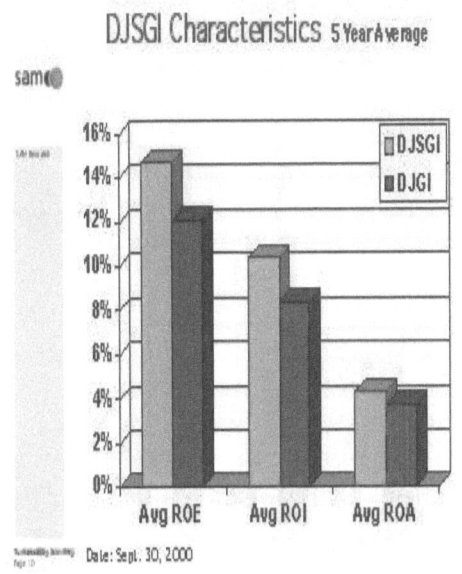

Leading DJSI companies must qualify in the following areas:

Strategy: Integrating long-term economic, environmental and social aspects in their business strategies while maintaining global competitiveness and brand reputation.

Financial: Meeting shareholders' demands for sound financial returns, long-term economic growth, open communication and transparent financial accounting.

Customer & Product: Fostering loyalty by prpviding a compelling client experience.

Governance and Stakeholder: Setting the highest standards of corporate governance, stakeholder engagement, corporate codes of conduct and public reporting.

Human: Managing human resources according to a sound, credible human capital plan.

"What is good for General Motors clients is good for General Motors" – Not the other way around

Sustainable, forward-thinking companies using double-bottom line business cases not only consider societal and environmental contributions as risk-reducing strategies or philanthropy, but as part of their core strategy for making money and maximizing their economic performance.

Even in the poorest markets, according to C.K. Prahalad: *"by stimulating commerce and development at the bottom of the economic pyramid, multinationals could radically improve the lives of billions of people and create a more stable, less dangerous world. Achieving this goal does not require multinational corporations to spearhead social-development initiatives for charitable purposes. They need only to act in their self-interest. Because these markets are in the earliest stages of economic development, revenue growth for multinationals can be extremely rapid. MNCs can also lower costs, not only through low-labor cost, but by transferring operating efficiencies and innovations to serve their existing operations in developed markets."* (Prahalad & Hammond, Serving the world's poor, profitably, 2002)

The first step in understanding business is focusing in the sources of wealth creation outside the organization and understanding the way clients and markets benefit and make a profit from products and services.

No company can do well when its customers and clients don't. No product or service can consistently make money by harming or impoverishing their end users and consumers.[24]

As obvious as this concept might seem to customers and entrepreneurs, it is usually hard to grasp for managers trained in the conventional "MBA" paradigm. (Mintzberg, 1994)

At the "seed" stage of business creation, most entrepreneurs tend to develop business plans, biased to one of their particular strengths – product, profits, clients, organization - to the detriment of other dimensions that might be critical for organization's survival and success..

Large, well-established companies, on the other hand, often produce an overpowering "planning inertia" that presses planners to "play safe" by repeating "proven" business models, products and services well

---

[24] Crime, substance abuse and prostitution can arguably be considered lucrative in spite of the damage they cause to their clients and customers health, wealth and life. Studies from opposite social and political perspectives on the incidence of criminal business on criminals' income and property value , coincide in demonstrate, however, that these type of businesses obtain also diminishing returns in terms of income and health that keep them failing at the highest rates and their workers struggling in the lowest income brackets. (Gladwell, 2002) (Sowell, 2006) (Ehrenreich, 2001)

past their prime[25] instead of engaging in vital but apparently riskier innovation[26].

Peter Drucker pointed out that the biggest risk for business was to keep social realities as "externalities[27]" out of the business formula, because *"the social dimension is a survival dimension. The enterprise exists in a society and an economy. Within an institution, one tends to assume that the institution exists in vacuum. And managers inevitably look at their business from the inside"*. (Drucker, The new realities, 1988, p. 37)

The first step in finding "the business" of our business, then, is to *look outward*, outside the organization, toward its clients, market, value chain, business ecosystem and environment and ask the simple question: *"how can we ensure that our customers and clients will be able to continue purchasing, using and recommending our products and services?'*.

A double-bottom line business case

Following Roger Kaufman's *Organizational Elements Methodology – OEM-* (Kaufman, 2006), we start developing our double-bottom line business case by identifying three levels of results: we must first *define and quantify the benefits of their project for the client, market and community* (Mega level), then *define products and services* (Micro

---

[25] Peter Drucker masterfully summarized this Fortune 500s "planning inertia" by coining the aphorism *"God condemns to 40 years of success those organizations he wants to destroy"* (Drucker, The new realities, 1988)

[26] Conventional marketing has done a lot of harm by **warning against** introducing "new products to new markets" (traditionally called "the death square"). A review of a century of business history (Chandler, 1990) shows that **successful companies do exactly that: introducing new products to new markets**. GM –on the other hand- followed that fateful "conventional wisdom" advice shelving the electric cars, ignoring hybrids and waiting for 150 dollars oil to react.

[27] According to conventional accounting doctrine, "an externality occurs when an economic activity causes external costs or external benefits to third party stakeholders who cannot directly affect an economic transaction" (Bernstein & Wilde, 2000) Treating social realities as "externalities" unrelated to business results turns profits into "pennies from heaven"

level) that our organization –and its allies in the value chain[28] and the business ecosystem[29]- must deliver to its clients in order produce those benefits for them and finally *define how the organization will, in turn, capture a part of those benefits as revenues* (Macro level), thus establishing a solid revenue stream for the new business.

In helping entrepreneurs and investors to plan new business at the "seed" stage or to define performance improvement projects for existing ones at the Sonora Institute of Technology (ITSON) business incubation program[30], we use a three-layered business case tool.

To illustrate the methodology, we will use as an example an actual three-level, double bottom line business case we developed for a business intelligence center in Mexico.

Trans-Pacific Center –TPC- is a spinoff of the Sonora Institute of Technology that enables Sonoran companies to export to Asian and Pacific coast markets –included the West Coast of the United States- by providing market research and reports, professional assistance on export-import procedures, coordinating support from other specialized

---

[28] We combined Geary Rummler's concept of "value creation chain" that with Kaufman's OEM to clarify how products and services from different organizations better coordinate and synergize to effectively add value to clients, clients' clients, investors and the market in general. We explain the basics of such methodology ahead in this article. A more thorough description can be found in Rummler's two books (Rummler & Brache, Improving performance: how to manage the white space in the organization chart, 1995) and (Rummler, Serious performance consulting, 2004) as well as in Brethower's Performance Analysis (Brethower, 2007), Maria Mallott's Paradox of Organizational Change (Mallott, 2003), Adam Afuah's Business Models (Afuah, 2004) and my last book, Capital Intelectual (Bernardez, Capital Intelectual: creacion de valor en la sociedad del conocimiento, 2008)

[29] A business ecosystem can be defined as a systemic integration of different organizations that support each other playing complementary roles in order to serve a market or community. (Iansiti & Levien, 2004). According to Bernardez, business ecosystems can –and must- be assessed as social capital and cooperatively engineered among multiple partnering organizations using this business case methodology and Geary Rummler's and Dale Brethower's Anatomy Of Performance –AOP- design tools . (Bernardez, Capital Intelectual: creacion de valor en la sociedad del conocimiento, 2008)

[30] More information about this unique PhD program at the Sonora Institute of Technology (ITSON) that graduates business instead of individuals can be found at www.piionline.org and in (ITSON - Sonora Institute of Technology, 2007) and (Rodriguez Villanueva & Guerra-Lopez, 2005)

Sonoran centers –such as laboratories specialized in food and animal health testing to meet international standards- and –last but not least- by helping local businesses find international clients that are the "right match" to each exporting company's competences and competitive advantage.

Table 1: Double bottom-line business case (Bernardez, Presupuesto Mega, 2007)

BUSINESS INTELLIGENCE CENTER (TRANS PACIFIC)						
US Dollars						
MEGA 'TOP LINE"	2007-2011	2007	2008	2009	2010	2011
Direct jobs	90	10	15	20	20	25
Ratio Indirect jobs/Direct Jobs		0.5	0.5	0.5	0.4	0.4
Indirect jobas	200	20	30	40	50	60
Annual average income Direct Jobs	5,362	5,365	5,368	5,371	5,374	5,377
Annual average income Indirect Jobs	5,362	5,365	5,368	5,371	5,374	5,377
Direct Jobs "ripple effect" revenue	483,495	53,650	80,520	107,420	107,480	134,425
Indirect Jobs "ripple effect" revenue	1,074,500	107,300	161,040	214,840	268,700	322,620
Tax revenue for State and Community	303	9,090	13,635	18,180	21,210	25,755
Exports revenue	47,019		10,909	11,454	12,027	12,629
MEGA RESULTS	1,605,014	170,040	266,104	351,894	409,417	495,429
MACRO "TOP LINE"	2007-2011	2007	2008	2009	2010	2011
Products & services revenue						
Research		3,600	4,200	4,800	5,400	6,000
Business plans		3,000	3,500	4,000	4,500	5,000
Special projects		2,500	3,750	3,750	6,250	7,500
Stages & exchange						
Other ITSON services (non BIC programs)		10,022	13,559	21,232	29,850	30,669
MACRO RESULTS	173,082	19,122	25,009	33,782	46,000	49,169
DOUBLE TOP LINE (MEGA+MACRO)	1,778,096	189,162	291,113	385,676	455,417	544,598
MICRO	2007-2011	2007	2008	2009	2010	2011
Products & services delivered						
Research	80	12	14	16	18	20
Business plans	80	12	14	16	18	20
Special projects	95	10	15	15	25	30
Stages & exchange	10	2	2	2	2	2
Other ITSON products & services	300	40	50	60	70	80
TOTAL MICRO OUTPUT	565	76	95	109	133	152
COST	2007-2011	2007	2008	2009	2010	2011
Initial investment	45,454	45,454				
Non-ITSON financial support	1,500	1,500	1,500	1,500	1,500	1,500
Licenses	1,000	1,000				
IT equipment	2,000	2,000				
Operational costs		3,600	3,602	3,604	3,606	3,608
Stages & exchanges	2,250	4,500	4,500	4,500	4,500	4,500
TOTAL COST	90,474	58,054	8,102	8,104	8,106	8,108
CONVENTIONAL BOTTOM LINE		-58,054	-41,147	-32,376	-20,160	-16,993
DOUBLE BOTTOM LINE		131,108	181,060	283,009	377,570	447,309
CONVENTIONAL ROI (MACRO/COST)		-1.28	-0.91	-0.71	-0.44	-0.37
SOCIAL ROI (MEGA+MACRO/COST)		2.88	3.98	6.23	8.31	9.84

By measuring the social and market impact of the *Trans-Pacific* business intelligence center over a five-year period, this three-level business case helps entrepreneurs make their case in order to obtain government and community support, attract angel investors, inspire staff and align all that the new organization produces and delivers with external, value-adding results.

The "top line" presents three different and related tiers:

A *Mega "top line"* –shown in Table 2-, reflects benefits and revenue generated by Trans Pacific business intelligence center to its clients, market and community in Sonora, Mexico: direct and indirect jobs created, direct and indirect jobs "ripple effect[31]" revenue for the local market, exports revenue for the local companies assisted by the TPC and tax revenue for government and community derived from TPC's clients increased business.

Table 2: Mega "top line"

BUSINESS INTELLIGENCE CENTER (TRANS PACIFIC)

US Dollars						
MEGA 'TOP LINE"	2007-2011	2007	2008	2009	2010	2011
Direct jobs	90	10	15	20	20	25
Ratio Indirect jobs/Direct Jobs		0.5	0.5	0.5	0.4	0.4
Indirect jobas	200	20	30	40	50	60
Annual average income Direct Jobs	5,362	5,365	5,368	5,371	5,374	5,377
Annual average income Indirect Jobs	5,362	5,365	5,368	5,371	5,374	5,377
Direct Jobs "ripple effect" revenue	483,495	53,650	80,520	107,420	107,480	134,425
Indirect Jobs "ripple effect" revenue	1,074,500	107,300	161,040	214,840	268,700	322,620
Tax revenue for State and Community	303	9,090	13,635	18,180	21,210	25,755
Exports revenue	47,019		10,909	11,454	12,027	12,629
MEGA RESULTS	1,605,014	170,040	266,104	351,894	409,417	495,429

This "social" top line serves two critical purposes: it indicates, in the first place, whether TPC clients will be able to pay for TPC services, providing a realistic estimate of its actual "market size", based not on "wishful thinking" or projecting generic demographics, but on the level of business to be generated by specific, targeted TPC clients' thanks to TPC's products and services.

---

[31] Based on previous research, we estimate "ripple effect" ratios on the local economy as a function of direct and indirect jobs' spending and of spending (and revenue) done by other business supplying the new ventures.

Additionally, this "Mega" top line shows local investors –"profit" (such as angel or venture capital) or "nonprofit" (such as local and state government)- the specific return on their investment in the TPC project.

A Micro "top line" –shown in Table 3-, reflects the products and services that both Trans Pacific Center and its allies[32] plan to deliver to its clients in order to produce the impacts and benefits described in the previous Mega-level top line results tier.

Table 3: Micro results "top line"

MICRO	2007-2011	2007	2008	2009	2010	2011
Products & services delivered						
Research	80	12	14	16	18	20
Business plans	80	12	14	16	18	20
Special projects	95	10	15	15	25	30
Stages & exchange	10	2	2	2	2	2
Other ITSON products & services	300	40	50	60	70	80
TOTAL MICRO OUTPUT	565	76	95	109	133	152

The Micro top line helps planners to define products and services *as means* to help clients achieve specific results –one of the key requirements for building a solid value proposition for the client- (Anderson, Narus, & van Rossum, March 2006).

Relating *products and services targets* shown in the Micro "top line" *to client's success and performance indicators* –shown in the Mega "top line"- sets the basis for realistic, competitive bidding and pricing –because measuring our company's products or services' benefits in terms of clients' specific results is critical to demonstrate the **Return to our clients on their Investment in our products and services.**

The Micro top line not only reflects products and services produced or delivered by TPC, but also those provided by allies, facilitating synergies, partnerships and alliances.

---

[32] Such as animal and food health labs and other providers of specific services for TPC –as tech support, languages, foreign service, etc.-

The Micro top line is expressed in product and services units –to be valued at the Macro "top line" level according to its ROI-based pricing[33]

A Macro "top line" –shown in Table 4- calculates the benefits for each company –in this case Trans-Pacific Center (TPC)- of providing products and services to the customers and clients identified in the Mega "top line" part of the chart.

Table 4: Macro "top line"

MACRO "TOP LINE"	2007-2011	2007	2008	2009	2010	2011
Products & services revenue						
Research		3,600	4,200	4,800	5,400	6,000
Business plans		3,000	3,500	4,000	4,500	5,000
Special projects		2,500	3,750	3,750	6,250	7,500
Stages & exchange						
Other ITSON services (non BIC programs)		10,022	13,559	21,232	29,850	30,669
MACRO RESULTS	173,082	19,122	25,009	33,782	46,000	49,169
DOUBLE TOP LINE (MEGA+MACRO)	1,778,096	189,162	291,113	385,676	455,417	544,598

Although most of these Macro-level benefits for TPC are monetized as revenues from products and services –or other forms of revenue such as licenses, franchise fees, copyrights, rent and other forms of intellectual, financial or structural capital rent – our Macro "top line" might include qualitative indicators of TLC's benefits, such as market share, market share growth, returns on investment (ROI), equity (ROE) or assets (ROA), client satisfaction and others.

Finally, following the aggregated costs column, *a double bottom-line* –shown in Table 5- reflects the net Social result (Mega + Macro-Costs) and the net Business result (Macro-Costs), thus allowing us to calculate the Social Return on Investment (Social ROI) and the Conventional Return on Investment (Conventional ROI), determining

---

[33] ROI-based pricing defines price as a function of the Return on Investment for the client shown by the relation Measurable Benefit for the Client (revenues, costs reduction) –Mega level- minus production and delivery Costs minus Profit Margin for the company. A ROI-based pricing revenue model helps TPC to focus in maximizing the actual value delivered as the key to maximize its own long-term (five year) revenue growth and profitability. (Bernardez, Tecnologia del desempeño humano, 2006) (Afuah, 2004)

the "break even" points for the organization and for its clients and market.

Table 5: Double bottom line

COST	2007-2011	2007	2008	2009	2010	2011
Initial investment	45,454	45,454				
Non-ITSON financial support	1,500	1,500	1,500	1,500	1,500	1,500
Licenses	1,000	1,000				
IT equipment	2,000	2,000				
Operational costs		3,600	3,602	3,604	3,606	3,608
Stages & exchanges	2,250	4,500	4,500	4,500	4,500	4,500
TOTAL COST	90,474	58,054	8,102	8,104	8,106	8,108
CONVENTIONAL BOTTOM LINE		-58,054	-41,147	-32,376	-20,160	-16,993
DOUBLE BOTTOM LINE		131,108	181,060	283,009	377,570	447,309
CONVENTIONAL ROI (MACRO/COST)		-1.28	-0.91	-0.71	-0.44	-0.37
SOCIAL ROI (MEGA+MACRO/COST)		2.88	3.98	6.23	8.31	9.84

The relation between Mega results –benefits for the client and its clients' clients, market and community- and these Macro results – benefits for our organization (in this case TPC)- reflect business long-term viability: if the Mega results –return to clients, market and community- are equal or below ($\leq$) the Macro results –benefits for the organization- it is likely that we have a non-sustainable business proposition[34].

*Macro results (organization's ROI) **are a function of Mega results** (clients, clients' clients, market and environment ROI)*[35]

---

[34] Imagine factoring the costs of healthcare created by a tobacco company to its clients and consumers, in terms of emphysema, cancer treatment, chronic illness and loss of quality of life –beside of life itself-. All those costs –ignored by conventional business cases- should be reflected in our Mega "top line" as *losses* or social costs, warning investors about the level of risk of such a business. Sooner or later, Mega (client) losses become Macro (business) losses, in the form of class action suits, compensations for damages, higher insurance. Take a look a the cases of tobacco, construction (asbestos), oil drilling (spillovers) and you might get a glimpse of what conventional business cases ignore at investors' own peril.

[35] That "nested", reciprocal ROI at the core of value creation is the "good" we talk about when we talk about "doing well by doing good". The same Adam Smith's "self-interest" principle that guides "creative destruction" at the macroeconomic level must guide the "double bottom line" at the microeconomic level. Reciprocal ROIs operate as a compass to survive.

Conversely, a non-profit organization –such as TPC in this case- might find useful setting Macro-revenue goals that ensure it will be able to self-finance its continuing value adding to clients and community. In our example, TPC managers must "return to the drawing board" to improve their Macro results –either through increasing revenues or by increasing public funding –government or angel capital- based on its proven social ROI.

Probing the "business" of the business case: engineering wealth

Although the double bottom-line business case methodology presented so far represents a a huge improvement in reducing investors', organizations' and communities' investment exposition to risk by disclosing and measuring benefits –and/or "external costs"- of a business project for clients, clients' clients and society, we still have to overcome the "Spanish objection" and make sure that the new business is as good in reality as it looks on paper.

In order to do so, we will use an extension of Geary Rummler's and Dale Brethower's *Anatomy of Performance –AOP-* process design methodology (Brethower, 2007) (Rummler & Brache, 1995) (Rummler, 2004) to disclose and analyze the value creation process proper and make sure that we follow Warren Buffett's two golden rules:

1. *"If you don't understand the business –be it a newspaper or a software firm- you cannot value the stock"*
2. *"Study prospects –and their competitors- in great detail. Look at real data –not at analysts' summaries-. Trust your own eyes."*
(Buffet, 2007) (Lowenstein, 1995)

Sharing with Geary, Dale and Roger Kaufman an even larger skepticism towards purely financial business cases, we went three steps beyond the "eye inspection" proposed by Buffett –still a key "due diligence" – and worked on a methodology that helps to magnify, design and test the "missing link" between the "strategic plan"[36] and the tactical and operational plans.

Let's break down the overall process in smaller, more specific steps:

Step 1: Identify benefits –build a three-tiered "top line" (Prahalad & Hamel, 1994)

> *"All value comes from outside the organization. Inside the organization there are only costs"*
> (Drucker, Management: tasks, responsibilities, practices, 1973)

Roger Kaufman used to say that the only way to guarantee an organization's long term survival and success is to make sure that the organization continues adding value to its clients and stakeholders.

Enron, World Comm and the current subprime mortgage crisis are stern reminders that *"if an organization doesn't add value to its clients, it is likely that it might end subtracting it"*. (Brethower, 2007)

Value doesn't grow out of spreadsheets, no matter how cleverly financial "Wizards of Oz" manipulate them: it comes from clients, markets and society.

Serious business planning can and must capture and relate all three levels of results –Mega, Macro and Micro-.

---

[36] Following Kaufman's advice, we consider defining the "top line" results based on a needs assessment the key components of strategy. Most companies label "strategic plan" to 100-pages, bloated documents which include tactical and even operational steps at the expense of solid foundation for their Mega, Macro and Micro results' definition.

During his consulting engagements, Peter Drucker used to hammer relentlessly CEOs and managers with a single, simple question: *"what is our business?"* (Drucker, 1973)
That question helped business leaders to continually make sure that the organization was adding value to clients and customers as a first requirement for its survival and prosperity.

In his seminal 1994 book *"Competing for the future"*, C.K. Prahalad went a step beyond Drucker, asking strategic business planners to look five to ten years ahead into the future, beyond current or past success formulas, in order to discover and seize the sources of wealth –actual value delivered to external clients and society- this additional question: *"what markets and clients will we be serving in five to ten years?"* (Prahalad & Hamel, 1994)[37]

In order to answer Drucker's and Prahalad's key strategic questions, our double-bottom line business case "top line" must reflect the three levels of results: *benefits for the client, market, community* (Mega) expressed as measurable performance results, *products and services* (Micro) that our organization must deliver to produce those results and finally, *revenues and non-monetized benefits* that our organization will receive (Macro) from clients, market and community in exchange for those products and services in order to be able to continue providing them, survive and grow[38].

Table 6 shows examples and sources of the three levels of results.

---

[37]Back in 1994, Prahalad stressed that ***"the organizational transformation challenge faced by so many companies today is, in many cases, the direct result of their failure to reinvent their industries and regenerate their core strategy a decade ago."*** (Prahalad & Hamel, Competing for the future, 1994). 14 years later we can see how many decisions on that decade have triggered the decline of industry leaders such as AOP, Sun or Microsoft.

[38] This is a powerful reason for non-profits to utilize a double-bottom line accounting and business modeling system. While for-profits tend to forget the client as their primary commitment and source of value(Mega), non-profits tend to forget their revenue stream and returns (Macro) that are essential to *continue* doing good. As Senator Sam Rayburn said to those complaining of the time spent in campaigning: *"in order to be a good senator, you must first continue being a senator"*

Table 6: Mega, Macro and Micro results in a "nutshell"

Performance level	Primary beneficiaries	Measured by –Key Performance Indicators (KPI)-
Mega	Society, economy	Roger Kaufman's Minimal Ideal Vision indicators (Kaufman, 2006)
	Clients	
	Clients' clients	Social indicators (jobs, ripple effect)
	Stockholders, investors	
		Social ecosystem
	Market	
		Market growth
		Direct and indirect client's performance (revenue, profit, productivity, cost savings)
Micro	End users	Products
	Clients	Services
		Delivery
Macro	Revenue for our organization and its	Revenue

stakeholders	Profit
	Market share
	Positioning, differentiation
	Attractiveness to financial, human and intellectual capital

Think Mega: searching for the business of business

Business propositions may start from very different points: they can come from a passion for a product, service or technology –like General Electric, Starbucks, Apple-; from the intellectual chemistry of a talented team –think of 3M, Hewlett-Packard, Google- or from specific core competencies –as in the cases of UPS, Dell, Amazon or eBay-. Business ideas can be revealed through flashes of rapturous inspiration –think of Mozart, Disney, MTV, Scotch™ tape, Alexander Fleming- or be the result of laborious, systematic research and incremental improvement processes –as in the case of Edison, Marconi, Ford or Boeing-.

But no matter where we start from or how we got there, producing the first definition of our project can be as frightening and stressful as staring at a blank canvas unless we have direction and tools.

In our experience helping a few hundreds of new businesses at the "drawing board" stage, we found four useful sources of "inspiration":

1. *Starting from a common, Minimal Ideal Vision (MIV) of the future*, as proposed by Roger Kaufman. The Minimal Ideal Vision is a device based on collected research that identifies

the common characteristics of the world people all over the world desires for the children of the future. All stakeholders must agree and commit to a shared vision based on the MIV.

You may use the MIV indicators summarized in Table 7 to derive social indicators in different areas such as self-sufficiency, well-being, security, safety, health, education or sustainability. The MIV-based indicators in your business case, however, must focus on actual, specific clients, markets and communities targeted by your business model[39].

Table 7: Minimal Ideal Vision (Kaufman, 2006)

There will be no loss of life or elimination of the survival of any species required for human survival. There will be no reductions in levels of self-sufficiency, quality of life livelihood, or loss of property from any source including:	There is a "gap" between what it is and what it should be
✓ war and/or riot and/or terrorism	
✓ shelter	
✓ unintended human-caused changes to the environment including permanent destruction of the environment and/or rendering it non-renewable	
✓ murder, rape, or crimes of violence, robbery, or destruction to property	
✓ substance abuse	
✓ disease	
✓ pollution	
✓ starvation and/or malnutrition	

---

[39] The scope of the business case sample must be targeted to real clients in order to maximize control over its variables and thus, validity and reliability to make it replicable in a larger scale.

✓ child abuse
✓ partner/spouse abuse
✓ accidents, including transportation, home, and business/workplace.
✓ discrimination based on irrelevant variables including color, race, creed, sex, religion, national origin, age, location
✓ Poverty will not exist, and every woman and man will earn as least as much as it costs them to live unless they are progressing toward being self-sufficient and self-reliant
✓ No adult will be under the care, custody or control of another person, agency, or substance: all adult citizens will be self-sufficient and self-reliant as minimally indicated by their consumption being equal to or less than their production.
Consequences of the Basic Ideal Vision: Any and all organizations-- public and private--will contribute to the achievement and maintenance of this Basic Ideal Vision and will be funded and continued to the extent to which it meets its objectives and the Basic Ideal Vision is accomplished and maintained. *People will be responsible for what they use, do, and contribute and thus will not contribute to the reduction of any of the results identified in this basic Ideal Vision.*

2. *Check Social Progress Opportunities (SPO) in current research* is another approach, based on analyzing current research on social trends and changes in different societies.

   Changes in technology and society define conditions and requisites for wealth creation.

   Clients and communities define and assign value to Micro-level products and services according to their current (and foreseeable) societal and technological environments.

   Global warming –as an example- is a menace for the continued value of CO_2-producing products and services as well, as an opportunity of appreciation for "green" technologies.

You might want to check Table 8 summary of several studies on US, OECD and BOP countries to identify possible indicators for your Mega-"top line".

Table 8: Social Progress Opportunities (SPO) and Technology Trends and Breakthroughs (TTB) recent research summary

Microtrends (Penn)	Megatrends (Toffler)	Flatteners (Friedman)	BOP (Prahalad)	Small Business (Barreto, SBA, Bernardez)
1. Commuter couples	1. Wealth, not just profits	1. Globalized market, free trade	1. Price-performance	1. Planning. Not winging it
2. Extreme commuters	2. Three Waves, three worlds (First: Agrarian, Second: Industrial society, Third: Information society)	2. WWW goes public	2. Hybrid solutions	2. Challenge CW
3. Stay-at-home workers		3. Work Flow software	3. Scalable	3. Build niche, differentiation
4. Stained glass ceiling breakers		4. Open-Sourcing	4. Eliminate resource wastage	4. Avoid copying and "bench-marking"
5. Sun-haters		5. Outsourcing	5. Functionality options	5. Manage and position in ecosystem
6. Long attention spanners	3. Clash of speeds	6. Off shoring	6. Process innovation	
7. Second-home buyers	4. Leaders & laggards	7. Supply-chaining	7. Deskilling	6. Manage "stages" of development
8. Smart Child left behind	5. Inertia	8. In sourcing	8. Consumer education	
9. LAT couples (UK)	6. Hyper speed	9. In-forming	9. Hostile environments-ready	
10. French teetotalers	7. Synchronization	10. Digital, Mobile, Personal, Virtual	10. User-interfaces	
	8. Future of job		11. Accessibility	
			12. Rapid evolution	

We have briefly summarized several recent studies, covering social Micro-trends –such as Mark Penn's findings in US and

OECD[40] countries (Penn & Zalesne, 2007) , focusing on new market "niches"- together with still valid Alvin Toffler's classical and recent studies which focus on larger social trends and markets. (Toffler, 1984) (Toffler & Toffler, 2006)

Table 8 also includes Thomas Friedman's technology-inspired "flatteners" and globalization effects (Friedman T. , 2005) – another factor to keep in mind when our business case clients are international or our partners and supply chain global- as well as C.K. Prahalad's revealing case studies on Bottom of Pyramid (BOP)[41] (Prahalad C. .., 2005) emerging economies success factors.

Finally, we also considered research about small business development critical indicators, both for the US (Barreto & Wagman, 2007) and for Latin America and emerging economies around the world. (Bernardez, 2007) (Bernardez, 2008)

3. ***Consider Technology Trends and Breakthroughs (TTB).*** Studies and research on technological changes and technologies actual and potential impact on social or environmental indicators can provide –that we also summarized in Table 8- can provide another valuable source for useful Mega "top line "indicators

4. ***Review your own Customer Experience (CE).*** Following Thomas Watson Sr.'s [42] advice: *"everything starts with a client"* (Watson Jr., 1990) , we should remember that if our organizations, products and services are means to clients' ends,

---

[40] Organization for Economic Co-Operation and Development: established in 1961, this international organization includes 31 most developed countries in the world –US, Europe and some emerging economies such as China, Brazil, Mexico and Russia-

[41] Term coined by C.K. Prahalad to refer to the markets of 4 billion people living with under $ 1,500 yearly income (Prahalad & Hammond, Serving the world's poor, profitably, 2002)

[42] IBM's founder

we must also draw into actual customer experience –both our own or others'- in order to find another source of value indicators for our Mega "top line".

The impact of food and food experience in variables such as blood pressure, weight, bone density, dental health and psychological well-being (or stress) are translated into economic value in terms of costs of different treatments, food waste or delivery time.

Mega-level, social indicators such as health, survival and well-being can be easily monetized if we *calculate the associated costs*[43] of medical treatment, medicine, loss of productive time, lower educational achievement and *calculate them as savings* or benefits for the client.

Table 9 summarizes the findings of different studies on the impact of different kinds of meals and food experience on professional women and their children.

Table 9: Client experience – Prepared Meals Example - (Smith & Wheeler, 2002) (Bernardez, 2008)

Key atribute	Performance indicators[44]	Value provided to client
Health	Normal blood pressure	$ 300 medicine/year
	Healthy weight	$ 1,200 healthcare/year
	Normal Bone density	$ 500 risk insurance/year
	No caries, dental problems	$ 1,500 dental care/year
Fun	No guilt	$ 200 antidepressants/year
	Fun to serve, to eat	Quality of life
	Fun cooking	Family life

[43] What W. E. Deming called "the cost of non-quality" (Deming, 2000). Only when we calculate the cost of the problem can we estimate the value of the solution

[44] You can also use this Table to identity Micro-level indicators about the features and performance of your products and services. Keep in mind to start from the "high-level" Mega indicators, which *refer to client benefits rather than to product & services features.*

Time	Time to eat together and enjoy	$ 200 fast food/junk food
	Zero or less than 5 minutes preparation	Quality of life Children care?
Pricing	Saving waste	$ 5,000 waste/year
	Saving time	$ 500 gas, cabs, delivery
	Saving healthcare costs	$ 1,200 healthcare/year
Convenience	Saving ordering time Saving carry out time	$ 500 gas, cabs, delivery

Think of people, communities, clients and situations that matters the most to you and / or your existing organization. Think of those you know the most, learn more about them, step in their shoes, and experience reality from their standpoint.

Step 2: Define a business value proposition for your primary and secondary clients

Identifying Mega results is critical to define a value proposition for the client that answers two questions that check your business fundamentals:

1. How can we *guarantee our clients' continued success*?
2. How can we *make sure our clients will be able to continue choosing and paying* for your products or services?

But in order to build our double bottom-line business case, we must go further.

A true value proposition for our client must not just identify benefits or compare and find generic advantages over competing alternatives, but also *demonstrate measurable, monetized return on investment to the client* based on its own specific requirements. (Anderson, Narus, & van Rossum, March 2006)

In order to do so, such value proposition must address questions 1 and 2 and focus on those benefits that are not only advantages for our product or service, but to do so in terms that address our target clients' specific requirements.

Table 10 illustrates another example of business value proposition, addressing the needs of professional women with children in a Mexican fast-growing town.

Defining needs as *gaps between current and desired results* (Kaufman, 2006) , we determined that the causes of the gaps in nutrition, health and well-being of young professional mothers of two have a yearly cost to clients –mothers- and "clients' clients" –their children- of average $ 9,300 a year, aside from qualitative factors such as stress, guilt and deterioration of family life[45].

Table 10: Business case value proposition for the client (Anderson, Narus, & van Rossum, March 2006)

Client	Current situation	Desired situation	Gap (causes)	Mega: value for the client	Micro: Products or services to close gap	Macro:
Women, profession al, 20-45 years with children **N=12,000** **Target= 50% 6,000**	Poor diet, lack of time, "microwa ve diet", adverse health effects in client (women)	Balanced diet, more time for at home meals, enjoy preparati on Avoid	Lack of time Waste of costly ingredient s "Fast food" Fast eating	(Cost of the problem) Food waste: $ 5,000 year Dental care: $ 1,500 /	Consists of:: Organic, scientific diet prepared food Health menus online	Margin = 15% 1,050 x client x year x target (6,000) = $ 630,00 0 net

---

[45] Which in turn could also be monetized for a longer than 5-year return period, if we consider what research shows in terms of learning disabilities, family violence and other behavioral problems down that road. Monetization depends on (a) reliable research and (b) the length of our business case "time series". (Hamilton, 1994) (Menard, 2002)

and client's "clients" (children) Overweight High blood pressure Loss of bone density Dental problems	adverse health effects: Normal weight, blood pressure Normal bone density Good dental health	(stress) Bad eating habits Lack of nutritious ingredients Lack of cooking skills Stressful cooking experiences	year Healthcare woman: $ 600 Healthcare year 2 children: $ 1,200 Bone density-related incidents (fractures) over 5 years: $ 1,000 Total= $ 9,300 year	Cooking 101 and cooking workshops Additional cost (products, services + "premium" price) = $ 7,000 Client savings = $ 2,300 a year	revenue

In order to properly resolve the causes of the gaps, an exploration of existing research literature indicates that we must combine a mix of products —such as organic food, healthy and gourmet-quality preparations and different balanced diets for different age, gender and activity requirements- with services —such as online menus, "fun cooking" and "cooking for (and with) the family" experiences and health indicators progress self-assessment checklists-.

Implementing this business formula might require to combine the *core competencies* (Prahalad & Hamel, 1990 May-June) of different organizations or specialists in a partnership —think of a mix of *WebMD*[46], *Dr. Phil*[47], *Food channel*™[48] and chef *Rick Bayless*[49]- to

---

[46] Online service that provides end users general health information, URL: http://www.webmd.com/

[47] **Dr. Phil** is an American talk show hosted by Phil McGraw. McGraw offers advice in the form of "life strategies" from his experience as a clinical psychologist. URL: http://en.wikipedia.org/wiki/Dr._Phil_(TV_series)

produce the desired impact on Mega indicators and collect the Macro revenues.

Or perhaps the business planners may prefer to develop "from scratch" a new organization with all these core competencies to deliver this value proposition in "one stop".

Step 3: Explain "how" value will be created and delivered: value-creation process design (high level)

In order to make visible "the business" part of the business case, we use a multi-level flow chart, shown in Table 11:

Table 11: Value Creation flow: combining Kaufman's OEM with Rummler's AOP

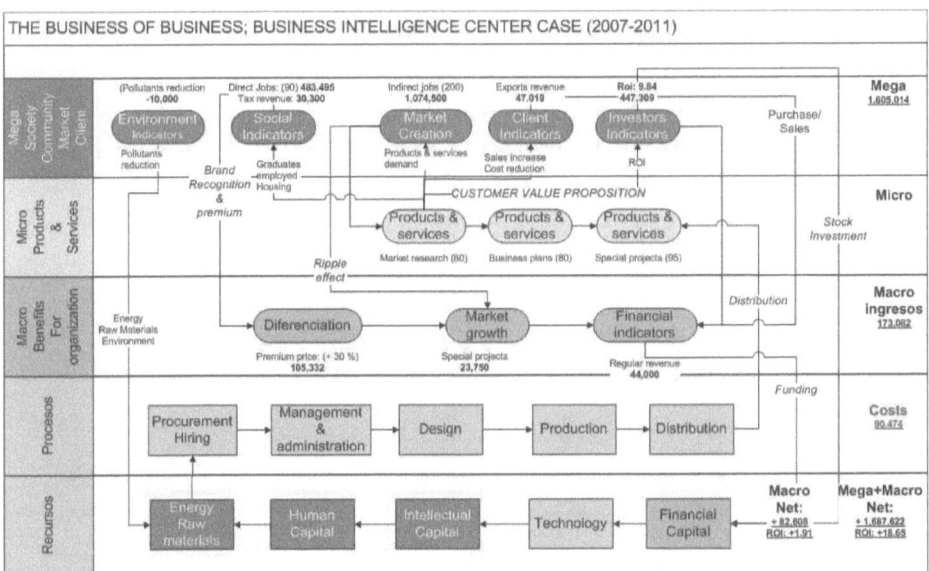

---

[48] Food Channel is part of Food Network, and online and cable TV network focused on cooking, food and nutrition information. URL: http://www.foodchannel.com/sections/3-home
[49] Chicago-based chef specialized in gourmet Mexican cooking. (Bayless & Bayless, 1987) URL: http://www.rickbayless.com/

This cross-functional[50] flowchart[51], based on Geary Rummler's and Dale Brethower's AOP model organized according to the levels of Kaufman's Organizational Elements Model –OEM: *Results* (*Mega, Macro* and *Micro*), *Activities or Processes* and finally *Resources*- has a "track" for each OEM component on and also a specific color[52] – darker for Mega, lighter for Macro, Micro and so on- for each symbol.

Each *flowchart symbol* represents a key component in the "business" of our business: the *value creation* process.

Mega, Macro and Micro results are specified and represented as *output symbols*.(○)

Flowchart's *connectors* reveal *how the business works* -the *dynamic* "wiring" of the business-, indicating how wealth flows from the Mega-level –clients, investors, community, environment- to the others.

Where value comes from: the Mega source

At the *Mega-level "flow line" track* –shown in the upper section of Table 12-, we can see TPC's main sources of wealth and value: environment indicators, direct jobs, indirect jobs (market creation), client indicators –such as sales revenues for exports and revenues for investors, costs savings –those investing in the export companies that our TPC is helping to find clients and position in foreign markets.

---

[50] And frequently "cross-organizational" flow chart, since different Micro products and services can be produced or delivered by different organizations partnering with ours. (Bernardez, Desempeño organizacional, 2007)

[51] A good introduction to flow charting can be found in Damelio's Process Mapping. (Damelio, 1996)

[52] Although the Mega level is "all Mega", there are Mega components and indicators at the Resource level: raw materials and energy coming from our environment, people who works in the production and delivery earning money in "direct jobs" that increase community's economy and market. Roger Kaufman uses to say "keep in mind that Mega is not in a far distant galaxy or statistics aggregates: it comes to work to your organization every day".

Table 12: Value creation source: Mega "top line" and flow line

US Dollars						
MEGA 'TOP LINE"	2007-2011	2007	2008	2009	2010	2011
Direct jobs	90	10	15	20	20	25
Ratio Indirect jobs/Direct Jobs		0.5	0.5	0.5	0.4	0.4
Indirect jobas	200	20	30	40	50	60
Annual average income Direct Jobs	5,362	5,365	5,368	5,371	5,374	5,377
Annual average income Indirect Jobs	5,362	5,365	5,368	5,371	5,374	5,377
Direct Jobs "ripple effect" revenue	483,495	53,650	80,520	107,420	107,480	134,425
Indirect Jobs "ripple effect" revenue	1,074,500	107,300	161,040	214,840	268,700	322,620
Tax revenue for State and Community	303	9,090	13,635	18,180	21,210	25,755
Exports revenue	47,019		10,909	11,454	12,027	12,629
MEGA RESULTS	1,605,014	170,040	266,104	351,894	409,417	495,429

Table 12 shows the alignment between the Mega, value-creation flowline and the Mega "top line" we use in the double-bottom line business case.

We can also see the revenue for TPC coming from client's –indicated as purchases and sales of TPC services- . The "virtuous circle" of value is here clearly depicted for all interested to see. Every connection and figure has to be backed for sound research data and can be tested and supported during the new business "pilot" run (incubation).

Delivering value proposition to the client: Micro to Mega

Once we established our clients' and clients' clients needs and identified Mega-level indicators, our next step is value-creation is to define which Micro-level products and services, organized as a value proposition, will have to be delivered by our organization –and its partners and allies to impact specific Mega indicators –as shown in Table 13-

Table 13: Delivering Value: Micro deliverables to respond to Mega needs

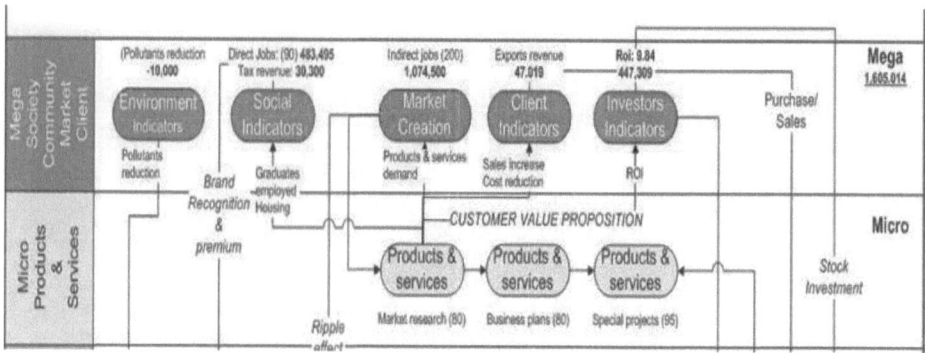

The *"upward-bound"* connectors going from Micro-level products and services –such as market research, business plans and special projects in our TPC case- to impact the five *Mega-level result **indicators** – environment (**pollution reduction**), social (**graduates employed**[53]), market (**demand of exportable products & services, indirect jobs**), client (**exports revenue**) and investors (**ROI**[54])*- display more clearly how our business *delivers* its customer value proposition.

Table 14 displays the process Micro-flow line aligned with the business case Micro "top line", detailing the expected Micro results in a five-year time line (2007-2011).

Table 14: Micro flow-line (value delivery) and Micro "top line" (business case)

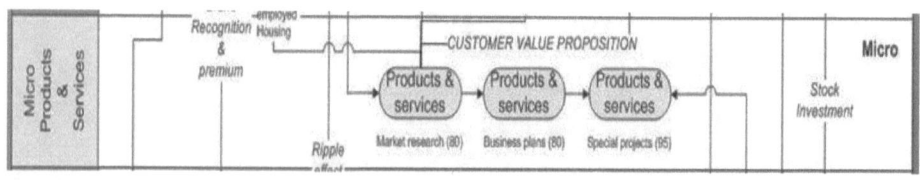

---

[53] As a university (ITSON) spinoff, TPC uses *"graduates employed by the new exporting ventures"* as a specific, Mega-level "Social Indicator"

[54] Return On Investment. (Bernardez, Tecnologia del desempeño humano, 2006) (Phillips, 1997) (Kearsley, 1982)

MICRO	2007-2011	2007	2008	2009	2010	2011
Products & services delivered						
Research	80	12	14	16	18	20
Business plans	80	12	14	16	18	20
Special projects	95	10	15	15	25	30
Stages & exchange	10	2	2	2	2	2
Other ITSON products & services	300	40	50	60	70	80
TOTAL MICRO OUTPUT	565	76	95	109	133	152

TPC project leaders can use the double bottom-line business case not only as a planning tool, but also for *managing the business*, following the phases *of planning, implementing* and *monitoring* as prescribed by Rummler's *Total Performance Management* (TPM) methodology. (Rummler & Brache, 1995)

Revenue flow: from Mega to Macro (and from Macro to Micro)

Table 15 three tiered flowchart shows the flow of benefits and revenue for the organization from the Mega level –environment, energy, raw resources, clients, clients' clients and investors- to the Macro level –in the form of purchases (clients), stock investment (investors), market growth (market creation) and brand recognition (social indicators).

Table 15: Revenue flow: from Mega to Macro

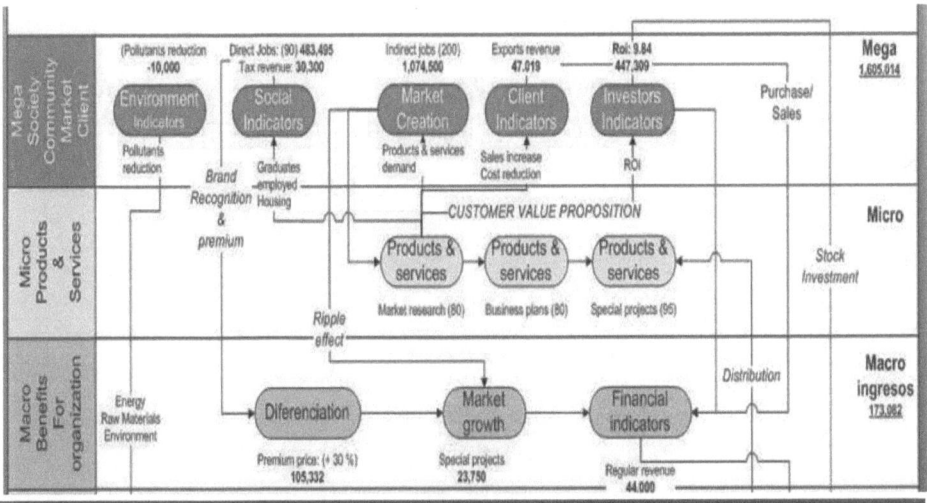

MACRO "TOP LINE"	2007-2011	2007	2008	2009	2010	2011
Products & services revenue						
Research		3,600	4,200	4,800	5,400	6,000
Business plans		3,000	3,500	4,000	4,500	5,000
Special projects		2,500	3,750	3,750	6,250	7,500
Stages & exchange						
Other ITSON services (non BIC programs)		10,022	13,559	21,232	29,850	30,669
MACRO RESULTS	173,082	19,122	25,009	33,782	46,000	49,169
DOUBLE TOP LINE (MEGA+MACRO)	1,778,096	189,162	291,113	385,676	455,417	544,598

Table 15 flow line also displays how other Mega factors –such as the environment- pay back to the Macro level in the form of cost savings for pollution reduction, but also to the Resources level, by providing a steady flow of energy and raw materials.

Inside the organization: Costs (Activities and Resources)

The value creation flow line goes now inside the organization –where planning, production, distribution and administration activities take place to transform resources into products and services-.

The flow line code colors keep track of the *origin*[55] of the resources: the Mega-level (environment and society) provides natural resources – such as energy and raw materials- , human capital (people) and intellectual capital (knowledge and know-how) –that "walk-in" every day from the community.

So does the financial capital provided by investors, collected at the Macro level and "distributed" through management and administrative support processes.

---

[55] Visualizing this "virtuous circle" of value creation and retribution helps to keep those inside the organization aware of the important fact that it is the external client and community what sustains the organization, making social focus not a matter of "charity" or philanthropy but what Adam Smith called strict, economic "self interest" (Smith A. , 1776)

Table 16: Inside the organization: Activities and Resources

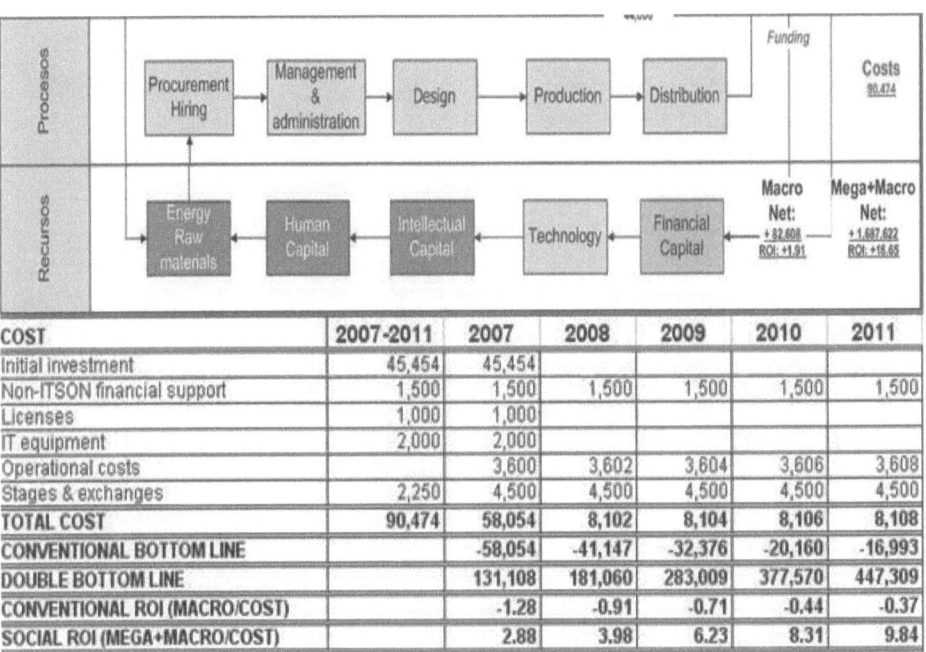

COST	2007-2011	2007	2008	2009	2010	2011
Initial investment	45,454	45,454				
Non-ITSON financial support	1,500	1,500	1,500	1,500	1,500	1,500
Licenses	1,000	1,000				
IT equipment	2,000	2,000				
Operational costs		3,600	3,602	3,604	3,606	3,608
Stages & exchanges	2,250	4,500	4,500	4,500	4,500	4,500
TOTAL COST	90,474	58,054	8,102	8,104	8,106	8,108
CONVENTIONAL BOTTOM LINE		-58,054	-41,147	-32,376	-20,160	-16,993
DOUBLE BOTTOM LINE		131,108	181,060	283,009	377,570	447,309
CONVENTIONAL ROI (MACRO/COST)		-1.28	-0.91	-0.71	-0.44	-0.37
SOCIAL ROI (MEGA+MACRO/COST)		2.88	3.98	6.23	8.31	9.84

Using a multi-level, cross-functional and even cross-organizational flowchart combined with our double bottom-line business case helps business planners and managers to track and manage the flow of value creation, connecting and aligning internal and organizational processes with results, suppliers and partners in the business ecosystem value chain that starts and ends outside the organization.

Using the double-bottom line business case and value creation flowchart for planning, management and evaluation helps prevent the "disconnects" and illuminate the "blind spots" between business planning, organizational design and execution that separate business success from failure.

Step 4: Zoom into the "value creation engine": define the client experience (where "rubber meets the road")

At its core, a business is as good as its weakest link: the point where products and services meet and impact clients, clients' clients, market and environment.

In the dynamic, global and nimble value chains that characterize the economy of the 21st century, the Value Creation Engine design should include the support of other partner companies that might provide complementary services or products forming a business "ecosystem" – such as *Apple* offering its *iPod* services in *Starbucks* stores or *FedEx* adding *Kinko's* stores mail, packaging and documents processing capabilities to its logistics to provide a "one stop", integrated client experience.

In order to understand the "business" of our business model, however, we must go a step further, zooming into the design of the "value creation engine" shown in Table 17.

Table 17 value creation process shows how the ***Ecosystem Synergies Design*** identifies and organizes those activities or processes that can be "farmed out" by our organization or become part of a partnership with allies "sharing" our clients' experience[56].

Client's experience constitutes the core of the "value creation engine" that delivers value to the customer and aligns all core processes to the Mega, Macro and Micro goals.

---

[56] A good example of this approach are partnerships such as Starbucks stores, wireless Internet providers and media companies that support the Starbucks' client experience. (Schultz & Jones Wang, 1997)

Table 17: Value creation engine and process

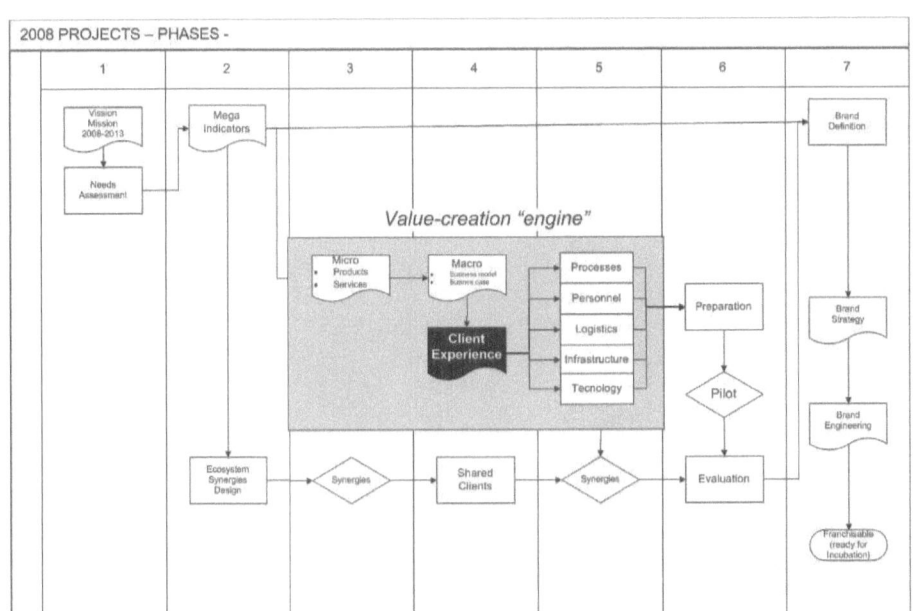

Following the business creation process sequence, Table 17 shows the previous steps in our business case: *–Vision and Mission design* (Phase 1), *Mega-level indicators* (Phase 2) , *Micro-level products and services* (Phase 3) and *Macro-level business case* and *revenue model* meeting the client in the *client experience* (Phase 4).

Investing in a business case without a clearly designed business experience is like buying a "concept car" that has no engine.

A client experience model helps planners and investors to pilot test the business case during the "incubation" phase[57], and as a sound, realistic starting point for designing effective processes, selecting adequate people, organizing our logistics, infrastructure and technology (Phase 5).

---

[57] Many startups "skip" the "incubation phase" only to discover that testing a business model in large scale under the pressure of competitive market conditions is much more expensive and leads to almost certain failure (Barrow, 2001) (Bernardez, Desempeño organizacional, 2007)

At its core –shown in Table 8- the Client Experience must define the standards that all key support factors –*technology, people, processes, logistics, and resources-* must meet in order to ensure consistent, reliable and successful implementation -instead of the other way around-.

Table 18: The Value Creation engine core: the Client Experience

The client experience applies to products as well as to services.

Clients differentiate and value products not just because of their technology or design, but because of their experience with them from the beginning –*marketing, information, purchasing* phases- along their life cycle –*reliability, performance over time* experiences- until the end –when *replacement, resale value, upgrading* performance can be decisive to make or break clients' loyalty-. All these factors are part of the client experience.

Mercedes-Benz customers switched to competitors like *Lexus* or *BMW* when *Mercedes* new technology proved unreliable, lowering the overall car quality ratings. *BMW series 5* customers switched to *Mercedes* and *Lexus* when they found the new electronic dashboard and hi-tech stereo commands unfriendly.

Otherwise loyal and enthusiastic customers might be turned off by products that have lower resale value –such as *Motorola's* frequently down prized *Rzr™* cell phones- or upgrades with downward-compatibility problems –such as *Microsoft's Vista-*

A client experience is a combination of (1) multiple *attributes* of the *relationship with products or services* that matter to the client and stages and (2) *steps* that the client goes through in his/her interaction and transactions with the company. Each step is what Jan Carlzon[58] called a "moment of truth" -a moment where the client actually compares what he/she gets with our business case "value proposition".

In order to generate loyalty to a brand, its client experience must also be *consistent, reliable* and easily *replicable.* (Smith & Wheeler, 2002). Clients must get used to get *what they expect the way they expected it* every time.

Building the business organization processes, logistics, competencies or administrative procedures around a clear definition of the client experience helps to avoid costly and frustrating process of "learning by doing" at the expense of client's loyalty and company's reputation[59].

Using the client experience matrix helps planners "tailor" products and services to the client, rather than the other way around.

Staff and franchisees must be able to get each new business unit running optimally in the fastest, most reliable way possible.

---

[58] CEO of SAS –Scandinavian Airlines-, author of the best-seller in quality of service"*The moment of truth*" (Carlzon, 1987)

[59] Even companies that hold a "quasi-monopoly" dominance of the market such as Microsoft in the Operative Systems, have paid a hefty price for testing their products on the real market, exposing users to miserable client experiences. The popularity of Microsoft's rivals such as Linux or Google stems from the resentment against the company and its products generated by those experiences.

A superior *client experience* is what distinguishes iconic brands such as *Starbucks, McDonalds, Whole Foods, Disney, Apple* or *Toyota* from the peloton of passing and failing "business experiences"[60], creating dedicated –even fanatic- customers that pay premium prices and "evangelize" other clients through "viral" marketing.

A well-designed client experience allows us to pilot test and evaluate our business model during the "incubation" (Phase 6) before "branding" and taking it to the market either by rising working capital in order to launch and manage a larger operation or by franchising the proven model to others (Phase 7).

A tested client experience is intellectual property of the highest value, franchisable and scalable. Companies such as *Starbucks, Disney, McDonalds* or *Wall-Mart* base their unique capacity for local and global growth in their winning combination of business models and proven, carefully designed client experiences.

In order to design a client experience we must define two basic components:

1. ***Experience attributes***: identify which are the key attributes that a client seeks and values in our products and services.

   Starbucks' clients value the design and relaxed environment of the European café that they find in the stores; the hi-speed, Wi-Fi connections and availability of sockets to connect their laptops for work or study, the variety of coffee flavors and the social and environmental commitment that the company shows in selecting and training its baristas and its Third World coffee-growers.

---

[60] According to US Small Business Administration – USBA-, each year 650,000 new small businesses (fewer than 500 employees) start in the US. Only 30 percent survives the second year. (USBA, 2005) (Bernardez, Desempeño organizacional, 2007)

Whole Foods' clients pay premium prices for the organic, pesticide-free products, as well as for the company's care for promoting small farmers, local sourcing, recycling bags and saving energy. Amazon customers love the one-click-away simplicity of the Web interface, the customer-centered suggestion tools, the one-day delivery capabilities and the easy return and guarantee policies. Think of your own company. Think as a client.

2. **Key stages and steps –*the "due process"*-:** identify the actual stages and steps of the interaction and transactions between the client and your company's products and services.

Amazon's clients are so enthusiastic about the simple, few steps and the straightforward, flexible processes to shop online in the store that no longer limit their purchases to books, but to an entire universe of merchandise. They are *Amazon*'s clients because of its sleek process that makes returns and refunds fast and straightforward, stimulating extra purchases.

Starbucks' clients love the social interaction and the relaxed pace of the processes. They value the quality time *Starbucks* offers for their stops at the store and the relaxation and enjoyment of the café product it provides.

Starbucks stores provide easy Internet access and a quiet, relaxed environment in which its customers can work, meet, read or listen to music and shop through Amazon.

Think of your own company, as a client: when do you have your first notice? How can we save clients' & company time? What are the key steps or transactions from the client's standpoint?

Once we sorted out attributes and steps, we can build a client experience matrix as shown in Table 19 for a bank branch.

Table 19: Client experience matrix (Banking example)

Client Experience	ATTRIBUTES						
	Fast services	Correctly managed interaction	Products and services knowledge	Knows client and his/her history	Provides adequate solutions	Fast problem-solving	Professional attitude and behavior
KEY STAGES AND STEPS — Open new account	One step , 5 minutes setup One form	Recognizes client Recalls all previous information	Explains, compares and tailors P&S to client needs	Recognizes client Follows up Explores needs	Offers solutions tailored to needs Implements and follows up	Helps client solve problems Takes charge	Saves time Achieves high client satisfaction Data 100% correct
Routine transactions	Under 5 minutes	One step, one person	Manages each product OK	Anticipates solutions	To client satisfaction	Saves client time	Saves time Zero mistakes
Financial planning and advice	One step assessment FAQ	Makes sure client understands plans and risks	Offers two best alternatives Compares	Tailors plans to client priorities	Brochure Calculator Simulation	Helps client get information	Minimizes risk Client understands
Credits and loans	Automatic scoring	Online pre-qualification	Knows client industry	Knows client history	Minimizes risk	Provides self-help to select	Idem + Meets S&L standards
Problem-solving	Routes correctly	Uses decision-making algorithms	Solves problems w/all products	Zero "old problems"	Cost-benefit-consequences	Gets it solved the 1st time	Zero recurrence Optimal solution
Information requests	Uses all job aids	Keeps human contact	Helps finding info Prints handout	Follows checklist QA	Saves time	Gets all info together	Instant < 5 min Zero recurrence
SUPPORT — Technology People Policies Processes	Usable Web EPPS tools One step process	Client information screens (EPSS)	Online products FAQ Training	Needs assessment online tool Commercial platform	FAQ system Simulator Web access	EPSS for problem solving Fast, reliable system	Screening Performance Appraisal Bonus Training

Support requirements such as technology, people skills and competencies, policies and processes derive from the requirements of the client experience. Each row shows the quality requirements to perform each step.

The lower row shows the support required to guarantee each service attribute.

Different job descriptions will share the same matrix, emphasizing in the overall quality of the client experience instead of "doing well their job".

Business products or services can be compared against competitors or alternatives using this chart as scorecard.

Only business models and brands that have a client experience "by design" [61] can be safely franchised. Without these specifications, each franchisee varies quality randomly, adversely affecting brand image and reputation.

Last but not least, checking the *Client Experience model* as value creation engine provides investors and evaluators a more clear understanding and estimate of the business of each business case.

When a double bottom line business case is for you

This is a summary checklist for you to check whether a double bottom line-business case is for your project:

A business case is for you if:	Yes	No
• Your company/project has a corporate social responsibility commitment		
• You are seeking public funds/ support		
• You are seeking "angel" capital investment		
• Sustainability is a critical factor for your project or organization		
• Market development is a priority		

• Your target market is a
  o New
  o Developing
  o BOP market		

• You (or your company) are
  o Starting a new business
  o A business incubator
  o Want to become a contributor to the new global realities		

• You want to show your client the full return of each $ spent on consulting investment		
• Your company is facing some of these		

---

[61] As opposed to unreliable, unpredictable client experiences "by chance" that are the hallmark of poor quality and poor service

challenges:
- o Inhospitable business environment
- o Loss of qualified workforce
- o Low engagement
- o Social liabilities
- o Community rejection / resistance

Five Easy Pieces

Five basic documents summarize the business case of our business:

1. Vision and Mission indicators
2. Double bottom line business case
3. Value creation flowchart
4. Value creation engine
5. Client experience matrix

Those five single-page pieces can convey a thorough understanding and a full picture of our business model to all stakeholders.

Some recommendations from experience

1. ***Keep it simple, avoid the triple, "quadruple, quintuple-top line"***: some authors separate "top lines" according with the kind of indicators: environment, social, cultural and so on. (Savitz & Weber, 2006) (Laszlo, 2003) (Holiday, Schmidheny, & Watts, 2002) (Henriques & Richardson, 2004)

   We beg to differ: the main purpose of our double bottom line-business case is to align and communicate business planning with business management and bookkeeping.

   We recommend keeping top lines simple, at no more of three levels (External Clients, Company, Products) so they provide a "big picture" view of the business in a single document.

Environmental, social or health variables can be included as separate item lines within a single Mega-level "top line" category.

2. **Stay practical: use sound research results, avoid "fads" and "snake oil":** a business case is as solid as the foundations and rationale for its numbers.

Be careful to use solid, validated research rather than opinion or commentary books.

Challenge the basic assumptions of your model with contrarian evidence until you make sure that your estimates are solid and realistic.

Play "straw man" and "devil's advocate" and make sure that all stakeholders find evidence as solid and reliable as you do.

Keep a human scale: avoid selecting Mega "top line" indicators that are too large to be sensitive –like regional GDP or per capita income of a city- for your project's scale and scope.

Think instead of your actual employees' income, health, housing or environmental indicators –variables that your organization can and must control- and about measuring indicators of the increased revenues or cost savings your products and services generated for actual clients –variables you must use to support your business proposition and calculate your prices-

Consider your initial business case as a research design: your variables must be under control, as well as the client and clients' client target population, so you can learn from the pilot test whether your assumptions were right or not, and your business models works or not, and, most importantly: "why".

Business people and scientists have opposite attitudes toward failure: for the conventional business person, failure "is not an option" because it usually involves bankruptcy and ends the venture. For scientists, failure is often the only reliable source of the knowledge required to succeed.

Under our incubation approach, our business case and client experience must be pilot-tested in a limited running before launching a full-scale operation and risking larger amounts of working capital.

This approach helps business to marry science in a way that has proven the most successful since Edison to our days. Make sure that all stakeholders from both backgrounds understand and share the model.

3. ***Mind your business: check the business case:*** your business case is not only a planning and modeling tool, but also an evaluation model, which should provide your organization with sound, field-tested evidence-based learning.

Additionally, the business case can (and should) be used by managers to run the business, trying to reach the targets of the three top lines and check the two bottom lines.

4. ***Mind your business (and your strengths and vocation): from "pasta factory" to "idea factory"***

When *Ray Kroc* –then a salesman of industrial mixers for restaurants- first met the *McDonalds* brothers diner in San Diego, California, he saw a business model were the restaurant-owners saw a successful business.

Kroc thought of franchising the model and putting McDonalds restaurants all over the country. The McDonalds were not interested and sold their name and "formula" for 2.7 million dollars in 1961 and continue operating their own restaurant until retiring some years later. (Kroc, 1977)

They never complained about their luck, not just because back in 1961, 2.7 million dollars guaranteed a good life after retirement, but because they were "hands-on" impresarios – those of the kind that enjoy operating a day-to-day business, like so many small "pa and ma" operations-.

Kroc, on the other hand was more attracted by the simplicity and replicability of the business idea than for McDonalds' hamburgers or San Diego. As a vendor and a salesman, he had always analyzed and compared business from a clients' perspective.

Kroc saw in the McDonalds brothers role models for future McDonalds' franchised restaurant operators: dedicated people –usually families with strong work ethics and a passion for cooking, cleaning and operating a store-.

Our double-bottom line and business model can be used both ways: to operate a single, successful business or to replicate it and franchise it to third parties. These are, however, quite different business approaches: the ownership approach focuses on produced capital - production and management- as value creators. The franchise approach focuses on intellectual capital –business models and engineering- as the "business" of the business.

You must ask yourself and your business partners about your strengths and preferences, as well as the potential of each approach before building your business case.

Summary

We hope these tools and models help you to better understand and evaluate the business of your business idea, as they have done for our entrepreneurs at the Sonora Institute of Technology.

A sound, verifiable and replicable business model is the cornerstone for success in the microeconomic world in which entrepreneurs, clients and managers work and live, and also provides a "reality check" and a risk-assessment tool for those operating in the macroeconomic world in which investors, economists and government operate.

The double-bottom line business case might help to increase new business' success rate by providing a sounder set of tools to midwife and incubate young enterprises.
It should also reduce investors and stakeholders' risk by focusing their attention into providing measurable value for clients and community, in the understanding that doing well and doing good are not contradictory, but mutually indispensable.

Bibliography

Afuah, A. (2004). *Business models: a strategic management approach.* New York:NY: McGraw-Hill.

Anderson, J. C., Narus, J. A., & van Rossum, W. (March 2006). Customer value proposition in business markets. *Harvard Business Review* .

Barreto, H. V., & Wagman, R. (2007). *The engine of America: The secrets to small business success from entrepreneurs who have made it.* New York, NY: John Wiley & sons.

Barrow, C. (2001). *Business incubators: a realist's guide to the world's new business accelerators.* West Sussex, UK: John Wiley & Sons.

Bayless, R., & Bayless, D. G. (1987). *Authentic Mexican: Regional cooking from the heart of Mexico.* New York:NY: Morrow.

Bernardez, M. (1996, October 12). *Rethinking the bottom line: Last frontier and first step for organizational change* . Retrieved September 4, 2007, from Performance Improvement Global Network,ISPI: http://www.pignc-ispi.com/articles/org-change/bernardez-bottom.htm

Bernardez, M. (2005). Achieving Business Success by Developing Clients and Community: Lessons from Leading Companies, Emerging Economies and a Nine Year Case Study . *Performance Improvement Quarterly, ISPI* , 37-55.

Bernardez, M. (2008). *Capital Intelectual: creacion de valor en la sociedad del conocimiento.* Chicago:IL: Global Business Press.

Bernardez, M. (2007). *Desempeño organizacional.* Chicago, IL: Global Business Press.

Bernardez, M. (2007). Presupuesto Mega. In ITSON, *Contribución de las Instituciones de Educación Superior a la Generación de Consecuencias Sociales Positivas* (pp. 56-95). Chicago, IL: Authorhouse.

Bernardez, M. (2006). *Tecnologia del desempeño humano.* Chicago, IL: Global Business Press.

Bernardez, M., Valdez, J. A., Santana, A., & Uribe, A. (2007). Coaching for new business creation. *International Journal of Coaching in Organizations* .

Bernstein, L. A., & Wilde, J. J. (2000). *Analysis of Financial statements, Fifht Edition.* New York: NY: McGraw-Hill.

Brethower, D. (2007). *Performance analysis: knowing what to do and how.* Amherst, MA: HRD Press.

Buffet, W. (2007). *Warren Buffett speaks: wit and wisdom from the world's greatest investor* . Hoboken:NJ: John Wiley & Sons.

Carlzon, J. (1987). *Moments of truth: new strategies for today's customer-driven economy.* New York, NY: Ballinger.

Chandler, A. D. (1990). *Scale and scope: the dynamics of industrial capitalism.* London: UK: Harvard University Press.

Damelio, R. (1996). *The basics of process mapping.* New York:NY: Productivity Press.

Davis, I. (2005, May 26). The Biggest Contract. *The Economist* , 87.

Deming, W. E. (2000). *The new economics for for Industry, Government, Education - 2nd Edition* . Cambridge:NA: MIT Press.

Drucker, P. (1985). *Innovation and entrepreneurship.* New York: Harper Business.

Drucker, P. (1973). *Management: tasks, responsibilities, practices.* New York: NY: Harper Business.

Drucker, P. (1988). *The new realities.* New York: NY: Harper Business.

Ehrenreich, B. (2001). *Nickel and dimed: On (not) getting by in America.* New York, NY: Henry Holt & Company.

Friedman, M. (1970, September 13). The social responsibility of business is to increase its profits. *The New York Times* .

Friedman, T. (2005). *The world is flat: a brief history of the twenty-first century.* New York: Farrar, Strauss & Giroux.

Gladwell, M. (2002). *The tipping point: how little things can make a big difference.* New York, NY: Back Bay Books.

Hamilton, J. D. (1994). *Time series analysis.* Princeton:NJ: Princeton University Press.

Holiday, C. O., Schmidheny, S., & Watts, P. (2002). *Walking the talk: the business case for sustainable development.* London: UK: Greenleaf.

Iansiti, M., & Levien, R. (2004). *The keystone advantage: what the new dynamics of business ecosystems mean for strategy, innovation and sustainability.* Boston, MA: Harvard Business Press.

International Bank for Reconstruction and Development / The World Bank. (1998). *Expanding the measure of wealth: Indicators of environmentally sustainable development.* Washington, DC: The World Bank.

ITSON - Sonora Institute of Technology. (2007). *Contribucion de las instituciones de educacionn superior a la generacion de consecuencias sociales positivas.* Chicago, IL: Authorhouse.

Kaufman, R. (2006). *Change, choices and consequences: a guide to Mega thinking and planning.* Amherst, MA: HRD Press.

Kaufman, R., Corrigan, R., & Johnson, D. (1969). Towards educational responsibity to society needs: a tentative utilitiy model. *Journal of socio-economic planning sciences* , 151-157.

Kaufman, R., Oakley-Browne, H., Watkins, R., & Leigh, D. (2003). *Strategic planning for success: aligning people, performance and payoffs.* San Francisco, CA: Jossey-Bass.

Kearsley, G. (1982). *Costs, benefits and productivity in training systems.* Reading:MA: Addison-Wesley.

Kroc, R. (1977). *Grinding it our: the making of MacDonald's.* Chicago, IL: Contemporary books.

Laszlo, C. (2003). *The sustainable company: how to create lasting value through social and environmental performance.* Washington, DC: Center for Economic Resources, Island Press.

Ley Toffler, B. (2003). *Final accounting: Ambition, greed and the fall of Arthur Andersen.* New York, NY: Broadway Books.

Lowenstein, R. (1995). *Buffet: the making of an American capitalist.* New York:NY: Broadway.

Mallott, M. (2003). *Paradox of Organizaitonal Change. Engineering Organizations with Behavioral Systems Analysis.* Reno, NV: Context Press.

Menard, S. (2002). *Longitudinal research.* Thousand Oaks: CA: Sage Publications.

Mintzberg, H. (1994). *Managers, not MBAs.* New York, NY: Berrett-Koehler Publishers.

Penn, M., & Zalesne, E. K. (2007). *Microtrends: the small forces behind tomorrow's big changes.* New york, NY: Hachette Book Group.

Phillips, J. (1997). *Return on Investment in training and performance improvement programs.* Houston: TX: Gulf Publishing.

Porter, M., & Kramer, M. R. (2002). The competitive advantage of corporate philantropy. *Harvard Business Review on Corporate Responsibility* , 27-65.

Prahalad, C. .. (2005). *The fortune at the bottom of the pyramid: erradicating poverty through profits.* Upper Saddle River, NJ: Pearson Publishing.

Prahalad, C., & Hamel, G. (1994). *Competing for the future.* Boston, MA: Harvard Business Press.

Prahalad, C., & Hamel, G. (1990 May-June). The Core Competency of the corporation. *Harvard Business Review* .

Prahalad, C., & Hammond, G. (2002). Serving the world's poor, profitably. *Harvard Business Review* , pp. 1-2.

Rodriguez Villanueva, G., & Guerra-Lopez, I. (2005). Educational Planning and Social Responsibility: Eleven Years of Mega Planning at the Sonora Institute of Technology (ITSON). *PIQ, Volume 18, Number 3* , 3-5.

Rummler, G. (2004). *Serious performance consulting.* Silver Spring, MD: ISPI/ASTD.

Rummler, G., & Brache, A. P. (1995). *Improving performance: how to manage the white space in the organization chart.* San Francisco: CA: Jossey-Bass.

Savitz, D., & Weber, K. (2006). *The triple bottom line.* San Francisco, CA: John Wiley & Sons.

Silbiger, S. (1993). *The ten day MBA.* New York: NY: Collins.

Smith, A. (1776). *The Wealth of Nations.* New York: NY: Bantam Classics.

Smith, S., & Wheeler, J. (2002). *Managing the customer experience: turning customers into advocates.* London, UK: Prentice-Hall.

Sowell, T. (2006). *Ever wonder why?* Stanford, CA: Hoover Institution Press.

Stiglitz, J. E. (2003). *The roaring nineties: a new history of the world's most prosperous decade.* New York, NY: W.W.Norton.

Toffler, A. (1984). *The Third Wave.* New York, NY: Bantam Books.

Toffler, A., & Toffler, H. (2006). *Revolutionary wealth: how it will be created and how it will change our lives.* new York, NY: Alfred Knopf.

USBA. (2005). Retrieved July 11, 2007, from United States Small Business Administration: http://www.census.gov/csd/susb/susbdyn.htm

Watson Jr., T. J. (1990). *Father, son & Co: My life at IBM and betond.* New York: NY: Bantam Books.

Yunus, M. (2003). *Banker to the poor: micro-lending leading the battle against world poverty.* New York, NY: Public Affairs.

Mariano Bernardez, PhD., CPT

Mariano Bernardez is an international consultant specialized in developing new businesses.

During a 30-year career as a consultant to Arthur Andersen, Andersen Consulting, United Nations Development Program, manager and CEO in charge of running new startups and consulting firms in Latin America, Europe and the United States, Bernardez has helped Fortune 500, small business, multinational corporations, governments and NGOs in bringing about new businesses and organizations.

Bernardez is research professor and director of the International Institute for Performance Improvement at the Sonora Institute of Technology (ITSON) in Mexico, a PhD and MBA program focused in improving social and organizational performance by developing new companies.

He is author of six books on the specialty and multiple articles in peer-reviewed and professional publications.

He has been Board Director at the International Society for Performance Improvement (ISPI) and is a frequent presenter at ISPI, ASTD, IFTDO and AMA international Conferences.

Surviving Performance Improvement "Solutions": Aligning Performance Improvement Interventions

By

Mariano Bernardez, PhD., CPT

"First, Do No Harm"
Hippocrates

According to a 1981 study, approximately one-third of patients' illnesses in a university hospital were caused by treatment. (Steel, German, Crescenzi, & Anderson, 1981)

With approximately 225,000 deaths per year, treatment-caused, *iatrogenic*[62] factors are the third leading cause of death in the United States – following heart disease and cancer. (Starfield, 2000)

According to 2000 statistics and research (Weingart, Ship, & Aronson, 2000) , treatment-caused deaths break down according the following leading causes:
- 12,000 - unnecessary surgery
- 7,000 – medication errors in hospitals
- 20,000 – other errors in hospitals
- 80,000 – infections in hospitals[63]
- 106,000 – non-error, negative effects of drugs

---

[62] The term *iatrogenesis* (from Greek: *Iatros:* physician) refer to adverse effects or complications caused by or resulting from medical treatment or advice.
[63] 2008 statistics show a growth in treatment-caused deaths and illnesses caused by hospital-breeded "superbugs" resistant to antibiotics. (Landro, 2008)

The most common causes of treatment-caused deaths are: (1) misdiagnosis; (2) drug interaction; (3) "nosocomial[64]" infections and (4) incorrect procedures.

"Solutions"-caused problems do not limit to healthcare centers: those familiar with home renovation can relate their experiences to the classic film *"Mr Blandings' builds his dream house"* (Potter, 1948),[65] where a New York publicist and his wife -longing to buy a country house in nearby Connecticut to escape Manhattan's crowded apartments- go through an ordeal of over costs, rework, conflicts with un-coordinated contractors, engineers and architects that end tearing down and rebuilding their new place in twice the expected time and at several times the original budget.

Unlike those classic Hollywood happy endings –the Blandings do get their dream house and live happily there ever after- , organizations engaging in ambitious "organizational change" or "performance improvement" programs often end experiencing recurrent, complex , systemic "hangovers" caused by multiple, uncoordinated and sometimes even conflicting solutions.

Did the CFO launched a *Management-By –Objectives –MBO-* initiative while the COO instituted a *Total Quality Management – TQM-* program? Did anybody read Deming's *TQM "11th commandment"*[66]? Did the CFO and COO know that those two "solutions" can be strongly antagonistic? Did they have time or tools to check the compatibility of such complex "solutions"?

Had the IT department purchased a costly and promising *Enterprise Resource Planning –ERP-* system *without knowing that* the CEO had

---

[64] Nosocomial: related to or acquired in a hospital or treatment center

[65] This classic comedy played by Carry Grant and Myrna Loy became a cult film, generating "tours" to the Blandings house and a sequel , *The Money Pit* (1986) with Tom Hanks

[66] One of W.F. Deming -TQM's founder- 14 points for TQM's implementation "11th: *Eliminate arbitrary numerical targets: Eliminate work standards that prescribe quotas for the work force and numerical goals for people in management"* (Deming, 2000)

just signed off an ambitious *merger agreement -M&A-* with a former rival with a non-compatible IT architecture?

Are several departments enthusiastically engaged in time-consuming, meeting-based training and organizational development initiatives to increase employees' engagement and improve climate?

Those and others are early warnings that our organization is descending into some sort of "transformational chaos" that will sooner or later become a larger problem on its own.

Changes in strategic plans, management, mergers and acquisitions and departmental initiatives combined with the multiple models, "solutions"-trained specialists and fads of a multi-billion dollar consulting industry end usually producing a level of systemic chaos that Gloria Gery aptly characterized as *"organizational flagellation"* (Gery, 1992) and is –in our experience as well as probably in yours- one of the major sources of employees' turnover, demotivation and resistance to participation in "performance improvement" initiatives.

Check our "early signs" of solutions-caused, *"iatrogenic"* problems checklist in Table 1. Consider each "Yes" a "flag" for a potentially serious problem.

Table 20: "Solutions" alignment problems early warning signals checklist

Indicators	Yes?
■ We are implementing a "(solution)" needs assessment	
■ We are improving a "(function)" performance improvement program	
■ We are improving *several* "(function)" performance improvement programs simultaneously	

- Complains about too many meetings for "improvement" initiatives

- PI projects at functional level

- Executive compensation tied to functional performance

- Vision and mission are general, rather philosophical statements, with only "soft" implementation

- No metrics are defined for vision and mission

- Each department or function has a written vision and mission statement

- Functions have strategic plans

- There is not a common definition of what is "strategic" (other than what the "(superior-level official)" wants

- Balanced scorecard is based on "adding up" functional goals

- Budget is based on "adding up" functional budgets and plans

- People complains about too much time in "improvement" or "change" projects at works' expense

As "solutions" vendors –deceitfully self-introduced as solution-"consultants[67]"- push to sell their "solutions" to companies' functional

---

[67] "Consultant" has increasingly become a "code word" for "sales person" of a "solution" rather than what it meant in the management or business associated to the likes of Peter Drucker, Roger Kaufman, C.K. Prahalad or Geary Rummler. Rummler titled appropriately his

areas –Human Resources, IT, Finance, Marketing and Sales being the most prolific buyers- that in turn "pull" from inside for their functional priorities: their climate surveys, ERPs, MBOs and CRMs , a flurry of unconnected and frequently conflictive "change" or "improvement" initiatives takes place in the organization[68].

Confusing "performance improvement" with "solutions implementation" often fails because this approach takes for granted that the mere "lack" of a given "solution" or resource –MBO, ERP, CRM- is a genuine organizational "need" or gap in organizational results. When systemic factors that cause the original problem are ignored, "solutions" consultants operate like the subcontractors in the Blandings' house, creating newly-bred problems and causing systemic chaos.

Furthermore, the "solutions" approach increases the chances of focusing needs assessment and interventions in optimizing subsystems -such as sales or financial performance- at the expense of organizational performance and external clients, investors' o societal interests.

Without a shared focus on organizational and external clients' results (Kaufman, Educational System Planning, 1972)–rather than functional efficiencies-, each new functional solution can become a nightmare for some other area or the entire organization.

Think of IT defining "spam" or Internet utilization automatic filters in a university with thousands of online educational users based on purely administrative criteria. Think –conversely-of an enthusiastic distance education department in the same university launching a new,

latest book "Serious performance consulting" as an indictment of such unethical practices. (Rummler, Serious performance consulting, 2004)

[68] In a not infrequent case of our experience studying "dysfunctional" performance improvement initiatives, two functional areas launched their "own" Balanced Scorecard initiatives limited to their functional "silos" and had their own separate Visions and Missions – as if they were separate organizations-. Interestingly, they hired the same "solutions" vendor – which didn't find abnormal or unethical such situation-

bandwidth-hungry virtual classroom on the standard IT network at peak time.

Function-focused solutions also fail because they treat a systemic problem –such as organizational performance or behavior- with partial "fixes" that ignore systemic connections and interactions between sub-systems at their own peril.

An MBO-based incentives program focused on improving individual results might end rewarding behaviors and decisions that produce losses to the organization –such as maximizing mortgage sales at the expense of credit risk-. A unilateral effort in maximizing bank tellers' courtesy and cross-selling effort might end causing clients' complaints about slow service. An equally partial emphasis on "fast service" might also end causing the bank to send clients looking for financial advice straight to the nearest competitor.

The root cause of most "solutions-caused" problems is the lack of a systemic, companywide, comprehensive model for planning and managing interventions. Left to "solutions: vendors, performance improvement interventions tend to run out of control, like the contractors in the *Blandings'* dream house, increasing costs and rework, reducing the chances for actual improvement and –last but not least- making the organization's staff and clients' lives miserable.

Systemic analysis breakthrough: Performance Improvement models

From its early origins in the work of Kaufman (Kaufman, Corrigan, & Johnson, 1969) (Kaufman, 1972); Brethower and Rummler (Brethower, 1972) (Rummler & Brache, 1995) and Gilbert (Gilbert, 1978) , those in the *performance improvement* or *performance system* [69]fields developed a unique focus on systemic analysis and solution that emphasized in

---

[69] Also called more recently –and controversially- Human Performance Technology by the International Society for Performance Improvement (ISPI

a) Considering performance and behavior as *functions* of a larger context or performance system (Brethower, 1972)

b) *Defining "need" as a gap between current and desired results*, not as a "lack of" resources or as a subjective "want" (Kaufman, 2006)

c) *Analyzing how all different factors interacting in a performance system affect performance* and performer and affect each other instead of blaming the performer (Gilbert, 1978) (Rummler & Brache, 1995) and

d) *Considering not just the individual, job-level factors* (Gilbert, 1978), but processes, organization (Rummler, 2004) and societal context (Kaufman, 2006).

These are the good news. The bad news are that because all PI / HPT different models were developed independently and successively in response to the challenges of *different performance levels* –individual, organizational or strategic, societal performance- , they do not "fit" very well and tend to be used as "alternative" approaches rather than as complementary.

Like the sages in the *Sufi* tale, those using a single model usually fail to get the "whole picture" and reducing their chances of coming with a comprehensive solution, fall in the trap of multiple, disconnected and finally dysfunctional initiatives whose success is –at best- temporary.

Uncovering the elephant sitting in the living room: integrating three performance levels

Although there are multiple performance improvement models –more than 46 according to different studies (ISPI, 2006) (Bernardez, 2006) (Dean & Ripley, 1997) (Kaufman, Thiagarajan, & MacGillis, 1997) -, they can be classified in three main categories according with their focus and scope:

- *Individual performance models* – such as Gilbert's *Six Boxes* (Gilbert, 1978), Mager's *performance analysis algorithm* (Mager & Pipe, 1983) , or Spitzer's *context of work* (Spitzer, 1986) (Spitzer, 1995) –

- *Organizational performance models* – such as Rummler's *Anatomy of Performance – AOP -* (Rummler & Brache, 1995), Brethower's *Total Performance System –TPS-* (Brethower, 1972) , Tosti's & Carleton's *Organizational SCAN* (Vanguard Consulting Inc., 1996) (Carleton & Lineberry, 2004) or Langdon's *Language of Work* (Langdon, 1995) – and

- *Strategic, societal performance models* –such as Kaufman's *Organizational Elements Model - OEM -* (Kaufman, Corrigan, & Johnson, 1969) (Kaufman, 2006) –

Individual performance models

Individual performance models –such as Gilbert's classic Behavior *Engineering Model –BEM-* (shown in Table 2) - are quite helpful in understanding and optimizing performance at the job level –the level of the individual worker-.

Table 21: The behavior engineering model (Gilbert, 1978)

	S^d Information	R Instrumentation	S_x Motivation
E – Environmental supports	**Data** 1. Relevant and frequent feedback about the adequacy of performance 2. Descriptions of what is expected of performance 3. Clear and	**Instruments** 1. Tools and materials of work designed scientifically to match human factors	**Incentives** 1. Adequate financial incentives made contingent upon performance 2. Nonmonetary incentives made available 3. Career-

	relevant guides to adequate performance		development opportunities
P – Person's repertory of behavior	**Knowledge** 1. Scientifically designed training that matches the requirements of exemplary performance 2. Placement	**Capacity** 1. Flexible scheduling of performance to match peak capacity 2. Prosthesis 3. Physical shaping 4. Adaptation 5. Selection	**Motives** 1. Assessment of people's motives to work 2. Recruitment of people to match the realities of the situation

Organizational performance models

A few years later, Gilbert's former business partners, Geary Rummler and Dale Brethower, took the entire approach to performance analysis and improvement several steps further in the systemic direction, noticing that using Gilbert's *BEM* model frequently led to optimize individual workers' performance at the expense of process and organizational performance.

If each worker were allowed to "improve" his/her own activities at the job level based on Gilbert "Six Boxes", regardless of other co-workers working ahead, before or while collaborating in *a shared work process*, their *collective* performance would experience a noticeable setback –as it would happen if each rower in a coxed four were to row at his/her own pace and rhythm-.

Rummler's and Brethower's[70] *Anatomy Of Performance –AOP-* model started by envisioning three levels of performance –nested one into

---

[70] According to their own report, Geary Rummler and Dale Brethower started expanding and questioning the primitive BEM model during their years of research together, and after parting for decades –Rummler to consulting, Brethower to academia- developed two models AOP and TPS that were in essence variations of a common one. They re-baptized it Anatomy of Performance –AOP- and have been working later years associated at ITSON with AOP.

and under each other-: job level, process level and function level –as shown in Figure 1-

Figure 2: Three levels of performance (Rummler & Brache, 1995)

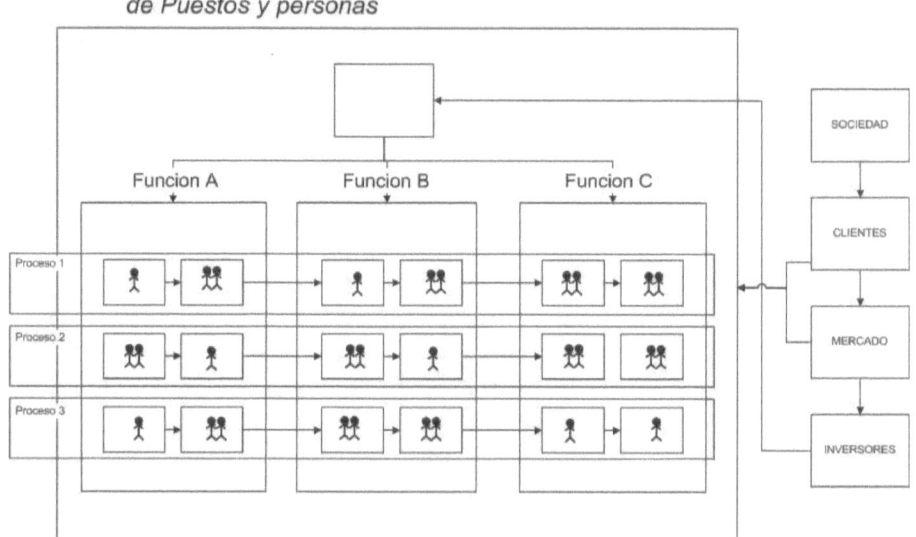

From (Bernardez, 2006)

Rummler's AOP model analyzes performance in three levels –*job, process, organization*- and at three levels of "performance needs": -*goals, design, and management*- considered from a performance management perspective.

Rummler & Brethower's matrix includes at the lowest level all key elements of Gilbert's BEM models, although not organized in *Six Boxes*[71].

---

[71] Although Gilbert's BEM formulation separates "environmental control" factors –see Table 2-, these are considered as part of the "job context" as in a "job description". The AOP model

Table 22: Geary Rummler's organizational "nine boxes" (Rummler & Brache, 1995)

		Performance needs		
		Goals	Design	Management
Performance level	Organization level	Organization objectives & indicators ■ Macro ■ Micro	Organization design ■ Macro ■ Micro	Organization management ■ Macro ■ Micro
	Process level	Process objectives & indicators	Process design	Process management
	Job level	Job and task objectives & indicators Resources levels & requirements	Job and task design Resources allocation system	Job and task management Resources management

Strategic, societal performance models

Although Rummler & Brethower's *AOP* and Tosti & Carleton's *SCAN* include references to the societal context considered as the "supra-system", their models do not pay such prior intense attention to societal performance as Roger Kaufman's *Organizational Elements' Model – OEM-* does.

Kaufman's *OEM* model –later reframed as part of his *Megaplanning* methodology- focuses on the planning process, particularly in

goes much further by differentiating "job conditions" such as these from process and organizational levels.

differentiating the true "strategic" part –represented by *Mega-level, societal results* driven by a *Minimal Ideal Vision* (MIV)[72] of the future- from "tactical" levels such as benefits for the organization – Macro- level goals such as revenue, market share or profit- and "operational" –which for Kaufman starts at the "outputs" (products or services) level or Micro-level and includes Activities[73]- and Resources –defined as inputs for Activities-.

Figure 3: Organizational Elements Model –OEM- (Kaufman, 2006)

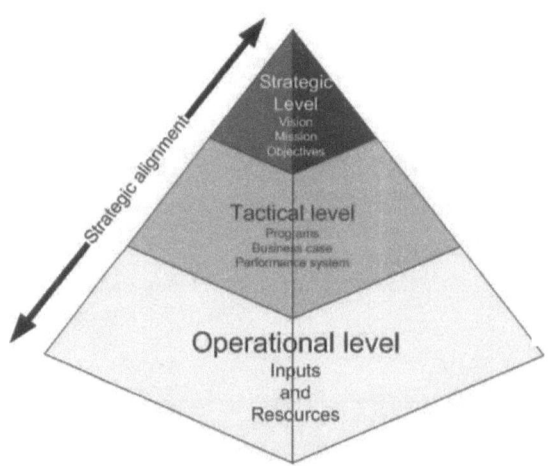

OEM levels

1. **Mega**
2. **Macro**
3. **Micro**
4. **Activities**
5. **Resources**

Kaufman's model focuses on establishing *vertical alignment* between strategic, tactical and operational results. Such alignment is –according to Kaufman- the only way to guarantee delivering actual value to external stakeholders, keeping the organization useful and focused.

---

[72] For a complete description of Kaufman's MIV see Kaufman's *Change, Choices and Consequences* (Kaufman, Change, choices and consequences: a guide to Mega thinking and planning, 2006) or Bernardez's *Tecnologia del Desempeno Humano* (Bernardez, 2006)
[73] Kaufman's"Activities" are a more general equivalent to what AOP defines as organization and process levels

Kaufman's *OEM* model provides a uniquely helpful guide to align all internal elements –after all, organizations are not ends unto themselves, but means to achieve results and produce value- and make sure that our sequence of definitions follows an "outside-in", "top-down" order rather than the other way around.

Using Kaufman's OEM as a *"builder's plumb"*-to continue with the home improvement analogy-, we can keep our performance improvement efforts honest and aligned with results and actual value for external stakeholders and the survival of the organization.

Integrating Kaufman's OEM, Rummler & Brethower's AOP and Gilbert BTE as shown in Table 4, we can connect and align the three levels of PI models and get a complete, systemic "blueprint" of what is involved in change and its probable impact in the organization's overall performance.

Table 23: Aligning models: multi-level framework

Level	Objectives Goals, standards & indicators	Design "How to", programs	Management Implementation, control
Societal/External(Mega) ■ **Community** ■ **Clients** ■ **Market** ■ **Suppliers** ■ **Value chain**	Mega objectives & indicators ■ Community ■ Market ■ People ■ Suppliers	Social & organizational plan strategic directions related to ■ ■ Community ■ Market ■ People	Social and regional management ■ Market ■ Policies ■ Regulations

			■ Suppliers	

Organization ■ **Macro (org. results)** ■ **Micro (products)**	Organization objectives & indicators ■ Macro ■ Micro	Organization design ■ Macro ■ Micro	Organization management ■ Macro ■ Micro
Processes ■ **Internal services**	Process objectives & indicators	Process design	Process management
People & resources ■ **"Six boxes"** ■ **Individual performer**	■ Job and task objectives & indicators ■ Resources levels & requirements	■ Job and task design ■ Resources allocation system	■ Job and task management ■ Resources management

Following a "falling water" path to establish PI intervention sequence

Applying principle (a) of all performance improvement models – adopting a *systemic* view - , we must start by positioning each intervention in one or more of *Table 4*'s 12 cells, according to the levels and steps where the intervention operates.

Following a *"falling water"* path, those PI interventions that operate at higher levels of the Framework –from upper left corner to the lowest right corner all across Table 4- have wider impact and affect the effectiveness of those at lower levels.

A change in our client definition, for example will affect our strategic programs. Changes in our strategic programs will subsequently affect strategic goals and rules, which, in turn, may trickle-down at the organizational goals-Macro level in our profit and loss statement.

When facing multiple performance improvement intervention, Table 4 will help us to:

a. Make sure that we start by implementing those interventions with the *highest and broadest systemic impact first*

b. Establish a *more effective sequence* of PI interventions,

c. *Explore the "downstream" effects* of each PI intervention in others and other non –PI programs and –last but not least-

d. *Keep our organizational budget under control.*

Conversely, if we start our performance improvement initiatives at lower levels, we may face unplanned "uphill battles" against overpowering, upper-level resistance factors that will increase the cost and effort required to achieve success and the risk of "systemic" failure –such as that of a costly IT system at the process level that ignores the strategic priorities of our external clients, wasting their time or delaying our organization's effective response to their demands.-

The guiding questions summarized in Table 5 help detect and prevent potential misalignment between levels (vertical alignment) and steps (horizontal alignment)

Table 24: Aligning Performance levels: Key questions

	Objectives	Design	Management
Societal External (Mega) o Community o Clients o Market o Suppliers o Value chain	❑ Is our organization adequate to meet the demands of new realities, clients, markets? ❑ Do we have a clear value proposition for clients and markets key for the future? ❑ Is our organization adding measurable value to the clients, market, communities it serves?	❑ Is organizational designed aligned with strategic vision and mission? ❑ is there a design of the future / desired organization? ❑ Are adequate indicators to measure the accomplishment and progress in achieving the strategic vision and mission defined and in place? ❑ Are the vision and future strategy adequately articulated as to de communicated	❑ Is management focused on developing and serving future clients? ❑ Has management a fluent and effective communication with clients, market, suppliers and community? ❑ Does management invest enough time in exploring future trends and changes in the market, client, supply chain, community?

	and guide implementatio n?		

Organization

- ○ **Macro (org. results)**
- ○ **Micro (products)**

❑ Are strategies correctly aligned communicat ed top-down, inside-out?

❑ Are our strategies compatible with our SWOTs?

❑ Do we have clearly defined products and standards for each level of results defined in our strategy?

❑ Do we have defined measurable

❑ Are all key organizational processes and functions clearly defined and implemented?

❑ Are all current functions adequate and adequately coordinated?

❑ Are products and services that link all functions consistent and adequate?

❑ Does the current organizational structure and functions adequately support the organizational strategy and

❑ Are each function goals clearly defined and coordinated with strategy and other functions?

❑ Is relevant function performance measured?

❑ Are resources adequately assigned ?

❑ Are interfaces between functions adequately coordinated?

results and standards at Mega, Macro and Micro levels?

performance?

❏ Are those results aligned and compatible with each other?

❏ Are our Mega results enough to support and sustain our Macro goals?

| **Processes**

Internal services | ❏ Are clear goals and standards defined for all key processes?

❏ Are those goals and standards aligned with organization's and client's requirement | ❏ Are the current processes the most effective and efficient to achieve the goals and meet the performance standards? | ❏ Are goals and standards for all key processes and sub processes clearly defined?

❏ Is process performance adequately measured?

❏ Are adequate resources for each key |

			process?
	s?		❑ Are process interfaces adequately coordinated?
People & resources ■ **"Six boxes"** ■ **Individual performer**	❑ Are Jobs goals and standards clearly defined and communicated to performers? ❑ Are Jobs goals and standards adequately aligned with processes' requirements?	❑ Are process requirements adequately supported by jobs and tasks involved? ❑ Are tasks and jobs' steps adequately sequenced? ❑ Are adequate policies and procedures in place? ❑ Are layout and technology adequate to support tasks and jobs?	❑ Do performers know & understand standards? ❑ Are resources and job design adequate? ❑ Are adequate incentives for meeting standards? ❑ Do performers know when they reach goals? ❑ Are they competent? ❑ Is job environment/ context adequate? ❑ Do performers have the

required
capacities?

Moving the elephant out of the living room: sequencing change interventions

> *"All change is disorienting.*
> *Too much change in too little time is destructive"*
> (Davis & McCallon, 1974)

We are not done yet in avoiding nasty collisions between "solutions".

Even combining all three performance levels *–external, organizational, individual-* we might discover that –like multiple contractors without a shared blueprint and plan- each improvement intervention might cancel, revert or force to redo previous ones, increasing costs as when carpets installed before piping or electric wiring must be removed and reinstalled.

Practical application in a university

We will put to use our integration framework in establishing a sequence for change and performance improvement interventions. We will use as an example a real case we solved using the tool.

Our client, a 12,000-student university was implementing multiple change and performance improvement initiatives –most started at the functional level- that were taking the entire organization to a grinding halt. People complained against "change" at all levels and yet, fought bitterly to give their own initiatives priority.

Along a 4-year period, a total of 10 different performance improvement initiatives were implemented in the following order:

Social and Organizational Performance Review

1. Each department launched *new educational programs* based on their experts' assessment of most valuable specialties and technological careers for the next decade.

2. A 10-year *IT infrastructure "master plan"* was launched to "unify the response" and "systematize" multiple departments' requirements "eliminating redundancies" and "setting common standards".

3. Due to inter-departmental conflicts during the first year of the IT plan, the university launched a cross-functional communications and team building initiative

4. As a result of the findings of the communications initiative, the university redesigned a common "value chain" linking educational programs with common goals.

5. In order to ensure the alignment between academic and administrative functions, both departments defined *Balanced Scorecards –BSC-* and *Strategy Maps* following Kaplan's methodology. (Kaplan & Norton, 1996) (Kaplan & Norton, 2004)

6. Based on the goals defined by the BSC process, the university launched a *market development plan* to assess the needs for future educational programs.

7. As a consequence of the conflict between the new educational programs -already designed in Step 1 of this list- and the Market Development study findings – unveiled on Step 6-, all university departments' decided to "update" their shared, strategic *Vision and Mission* based on Kaufman's OEM concepts.

8. Responding to the evidence of insufficient exchange of knowledge and know-how across disciplines, the university started a inter-disciplinary *Knowledge Management* initiative.

9. Based on the findings of both BSC and Communications programs, a cross-functional *Change Management initiative* was launched to facilitate the migration from the traditional culture to a new culture.

10. Responding to recommendations from the Change Management initiative, the university launched a *process reengineering* program in order to transition from a "functional" to a "process management" methodology.

The consequences of all this "zigzagging" decision-making process driven by reactions and "fixes" to unexpected consequences of each improvement step were dire.

At the time of our first assessment, most research professors were investing more time in multiple "performance improvement" initiatives than in their primary research or teaching jobs.

Morale was sinking, complaints were mounting and the Department heads and career directors felt confused and frustrated. One of the directors summarized the "change management" team feelings about the overall performance improvement process, at that time: *"We're deep into Alice in Wonderland's rabbit hole"*[74].

Like homeowners lost in a vast reform project without blueprints, the multiple "rabbit holes" created by each performance improvement "step" were delivering new systemic emergencies to address without any end in sight.

---

[74] Reference to Lewis Carroll novel's character, Alice, who in chasing her rabbit through a rabbit hole falls into a parallel world –Wonderland- a magic kingdom hidden behind Alicia's lookinglass whose inhabitants –the Mad Hatter, the Queen of Hearts- turn logic upside-down. (Carroll, 1865, 2000)

Back to reality: fixing the "lookinglass" with the PI interventions alignment tool

Based on the University newly defined Vision and Mission and its specific, measurable indicators focused on the university external clients' priorities, we helped the university management get out of their Alice's "rabbit hole" by revising and re-sequencing their multiple performance improvement initiatives.

We asked the team to organized the 10 interventions using using our *Table 4 multi-level, PI models alignment framework*, with the results shown in Table 6.

Table 25: Aligning performance improvement interventions - Example

Level	Objectives	Design	Management
	Goals, standards & indicators	"How to", programs	Implementation, control
Social External (Mega) ■ Communnity ■ Clients ■ Market ■ Suppliers ■ Value chain	*Vision and Missio n (1)*	*Market development plan (2)*	

Organization	Value chain redesign (3)	Balanced Scorecard (4)
■ Macro (org. results) ■ Micro (products)		
Processes ■ Internal services	Process reengineering (5)	Change Management (6) KM (7)
People & resources ■ "Six boxes" ■ Individual performer	IT infrastructur e (8)	Communications & Team-Building (9) New Educational Programs (10)

Colleges and other professional organizations established around specialized professional disciplines have a known tendency to create parallel, redundant planning and organization structures.

University departments tend to operate in relative isolation from each other, because –according to Daniels and Mathers'-:*"professionals go to their tasks alone; they gain skill from their own experience and the sharing of that experience with fellow professionals"* (Daniels & Mathers, 1997) .

This is often the reason why academics often fail to understand that *"reality"* -as Roger Kaufman uses to say- *"is not divided in disciplines or departments"*. (Kaufman, 2006)

Based on this new sequence defined in Table 6, we estimated the "gap" between the "trial and error" sequence followed originally and what would be a logical "top-down" sequence.

We asked the evaluation team to estimate the consequences of each "gap", monetizing the cost wherever possible. Table 7 shows the results.

Table 26: PI interventions misalignment: calculating the costs of "non-quality"

Performance Improvement Initiatives	How it happened "solutions"-focused	How it should be prioritized	Gap[75]	Impact and cost of the Gap
New Educational programs	1	10	+9	Redesign, Low enrollment **(50,000)**
IT infrastructure (five years)	2	8	+6	Repurchase, retrofit **($ 50,000)**
Communications & team building	3	9	+6	Downtime, **($ 40,000)** repetition (over cost)

---

[75] A + (plus) sign implies that the intervention was incorrectly anticipated to others. A – (minus) sign implies the intervention was incorrectly delayed.

Social and Organizational Performance Review

				($ 35,000)
Value chain redesign	4	3	(1)	Lost contracts (**$ 30,000**)
Balanced Scorecard (BSC)	5	4	(1)	Rework **($ 25,000)**
Market development plan	6	2	(4)	Opportunity cost **($ 50,000)**
Vision and Mission	7	1	(6)	All of the above, misalignment
Knowledge Management	8	7	(1)	Loss of information **($ 12,000)**
Change Management program	9	6	(3)	Resistance, conflicts **($ 20,000)**
Process reengineering	10	5	(5)	Rework **$ 30,000)**
Potential savings using the PI interventions alignment framework:				**($ 342,000)**

The root cause of the conflicts, rework and frustration ending in a $ 324,000 loss and –more importantly- an increased resistance to change, was the lack of an integrated, multi-level model to organize and prioritize different PI interventions putting the organization's results –"health"- ahead of those of the functional areas and their "solutions" consultants –"medicine"-.

Using this comprehensive, multi-level interventions organizer as a common framework allowed all "change advocates" championing specific "solutions" to identify and prevent systemic problems and re-organize the existing PI programs in an effective manner, instead of competing and fighting each other.

Conclusion

Improving organizational performance is too important to leave it to multiple "performance improvement models" or "solutions" consultants.

Failing to align and integrate multiple PI "solutions" not only has immediate, measurable and costly consequences, but the long-lasting effect of creating or reinforcing "change aversion" in the organization and its internal and external stakeholders.

Using a multi-level, organization and external clients-focused framework may help organizations and consultants to achieve better results with less pain and effort.

This -by the way-, is what "performance improvement" is all about.

Bibliography

Bernardez, M. (2006). *Tecnologia del desempeño humano*. Chicago, IL: Global Business Press.

Brethower, D. (1972). *Behavioral analysis in business and industry: a total performance system*. Kalamazoo, MI: Behaviordelia, Inc.

Brethower, D. (2007). *Performance analysis: knowing what to do and how*. Amherst, MA: HRD Press.

Carleton, J. R., & Lineberry, C. (2004). *Achieving post-merger success*. San Francisco, CA: John Wiley & Sons.

Carroll, L. (1865, 2000). *Alice's Adventures in Wonderland*. London, UK: Signet Classics.

Daniels, W. R., & Mathers, J. G. (1997). *Change-ABLE organization: Key management practices for speed and flexibility*. Mill Valley, CA: American Consulting & Training.

Davis, L. N., & McCallon, E. (1974). *Planning, conducting and evaluating workshops*. Austin, TX: Learning Concepts.

Dean, P., & Ripley, D. E. (1997). *Performance Improvement Pathfinders: Models for Organizational Learning Systems*. Washington, DC: ISPI.

Deming, W. E. (2000). *Out of the crisis*. Cambridge: MA: MIT Press.

Gery, G. (1992). Organizational flagellation. *CBT Directions*, 23-30.

Gilbert, T. F. (1978). *Human Competence: engineering worthy performance*. Amnherst, MA: HRD Press.

ISPI. (2006). *Handbook of Human Performance Technology (Third Edition, Pershing, R., edtor)*. San Francisco, CA: John Wiley & Sons.

Kaplan, R. S., & Norton, D. P. (2004). *Strategy Maps: Converting intangible assets into tangible outcomes*. Boston, MA: Harvard Business Press.

Kaplan, R. S., & Norton, D. P. (1996). *Translating Strategy into Action: The Balanced Scorecard*. Boston, MA: Harvard Business Press.

Kaufman, R. (2006). *Change, choices and consequences: a guide to Mega thinking and planning*. Amherst, MA: HRD Press.

Kaufman, R. (1972). *Educational System Planning.* Englewood Cliffs, NJ: Prentice-Hall.

Kaufman, R., Corrigan, R., & Johnson, D. (1969). Towards educational responsibity to society needs: a tentative utilitiy model. *Journal of socio-economic planning sciences* , 151-157.

Kaufman, R., Thiagarajan, S., & MacGillis, P. (1997). *THe Guidebook for Performance Improvement: Working with individuals and organizations.* San Francisco, CA: Jossey-Bass/Pfeiffer.

Landro, L. (17 de September de 2008). Rising foe defies hospital wars on "superbugs". *Wall Street Journal* , págs. D1-D6.

Langdon, D. (1995). *The new language of work.* Amherts, MA: HRD Press.

Mager, R., & Pipe, P. (1983). *Analyzing performance problems - or Your Rsally Oughta Wanna.* Atlanta, GA: Center for Effective Performance.

Potter, H. (Dirección). (1948). *Mr. Blandings Builds His Dream House* [Película].

Rummler, G. (2004). *Serious performance consulting.* Silver Spring, MD: ISPI/ASTD.

Rummler, G., & Brache, A. P. (1995). *Improving performance: how to manage the white space in the organization chart.* San Francisco: CA: Jossey-Bass.

Spitzer, D. (1986). *Improving individual performance.* Englewood Cliffs, NJ: Educational Technology Publications.

Spitzer, D. (1995). *Super-Motivation: A blueprint for energyzing your organization from top to bottom.* New York, NY: AMACOM.

Starfield, B. (2000). Is US health really the best in the world? *JAMA* , 284 (4): 483–5.

Steel, K., German, P. M., Crescenzi, c., & Anderson, J. (1981). Iatrogenic illness on a general medical service at a university hospital. *New England Journal of Medicine* , 304 (11): 638–42.

Vanguard Consulting Inc. (1996). *Organizational SCAN.* Unpublished manuscript: Vanguard Consulting Inc.

Weingart, S., Ship, A. N., & Aronson, M. D. (2000). Confidential clinician-reported surveillance of adverse events among medical inpatients. *Journal General of Internal Medicine* , 15 (7): 470–7.

Mariano Bernardez, PhD., CPT

Mariano Bernardez is an international consultant specialized in developing new businesses.

During a 30-year career as a consultant to Arthur Andersen, Andersen Consulting, United Nations Development Program, manager and CEO in charge of running new startups and consulting firms in Latin America, Europe and the United States, Bernardez has helped Fortune 500, small business, multinational corporations, governments and NGOs in bringing about new businesses and organizations.

Bernardez is research professor and director of the International Institute for Performance Improvement at the Sonora Institute of Technology (ITSON) in Mexico, a PhD and MBA program focused in improving social and organizational performance by developing new companies.

He is author of six books on the specialty and multiple articles in peer-reviewed and professional publications.

He has been Board Director at the International Society for Performance Improvement (ISPI) and is a frequent presenter at ISPI, ASTD, IFTDO and AMA international Conferences.

Are Performance Improvement Professionals Measurably Improving Performance? A look at what PIJ and PIQ have to say about the current use of evaluation and measurement in the field of performance improvement

By

Ingrid Guerra-Lopez, PhD. & Hillary N. Leigh

INTRODUCTION

The ability to prove that performance improvement professionals have made a measurable contribution to their clients and the field remains uncertain (Kaufman & Clark, 1999).

Clark & Estes (2000) noted that highly regarded research groups who surveyed performance improvement solutions found "a huge gap between what we think we accomplish and what scientific analyses say we accomplished." (p. 48).

Below are some of the findings cited by Clark and Estes (2000) based on the work of the National Academic of Sciences and the National Research Council and other independent research groups:

- Scientific studies of training found training interventions often leave participants worse off than before the training intervention (e.g. more confused, less able to remember important information; less able to use their work-related knowledge effectively

- More than half of organizational change initiatives are quickly abandoned

- Kirkpatrick's level 1 evaluation, the most commonly used method for evaluation, often gives about as much inaccurate information as it does accurate information, including the perception that the object of evaluation has helped, when if fact it has done quite the contrary

- Studies have shown that employee empowerment strategies have minimal success in some organizations, and have negative consequences in others. The more rigorous the evaluation, the less likely one is to find evidence of success.

- A myriad of studies have found no evidence that multimedia, internet, intranet training provide additional learning benefits beyond those already provided by traditional mediums such as human trainers or manuals.

- Studies indicate that one third of the feedback strategies employed in our field, does not improve performance, and another third make performance worse.

- Experiments that check for transfer of performance solutions show that while they work once, the almost never work on other organizational contexts. Since we don't evaluate solutions that may have worked for someone else in other organizational context, we remain ignorant of this failure to transfer

- Successful performance improvement strategies do exist, however, they are seldom integrated into our most popular performance solutions.

Clark and Estes,(2000) also argue that performance improvement professionals tend to "scientize" craft solutions by citing research and evaluation that is often irrelevant and/or poorly designed.

This could suggest a number of things, chiefly a) performance improvement professionals do not know how to integrate appropriate research and evaluation practices and findings into their work or; b) they do not want to integrate appropriate research and evaluation or; c) they neither know how to, nor care about integrating appropriate research and evaluation to their work, or d) they are unaware of the importance of integrating appropriate research and evaluation practices into their work.

This challenge is also faced by other fields closely related to performance improvement. For instance, field of applied behavior analysis (ABA), and organizational behavior management (OBM) have "also faced the challenge of extrapolating basic experimental research findings to the behavior of individuals at home, school, work, and in the community" (Culig, Dickinson, McGee, & Austin, 2005).

Dickinson (2000) cites various early studies (e.g. Andrasik, 1979; Frederiksen & Johnson, 1981; Frederiksen & Lovett, 1980; Hopkins & Sears, 1982; O'Hara, Johnson,& Beehr, 1985) that support the effectiveness of early organizational behavior management interventions, but also states that OBM requires increased measurement of social validity, cost/benefit analyses, and employee satisfaction/resistance; as well as the long-term effects of interventions.

Moreover, the author also cites Balcazar, Shupert, Daniels, Mawhinney, & Hopkins, 1989, who argue that there is a lack of "largescale interventions in which behavior principles are employed to change the 'cultural foundations' of an organization" (p. 36).

If performance improvement professionals are going to be taken seriously in the scientific and professional communities, they must be able to provide evidence of the rigor and seriousness with which they conduct their work.

Publications are artifacts that reflect what a field is about and where its members place their priorities.

As such, we want our publications to reflect the true intent, and hopefully actual practice, of the performance improvement field.

Social and Organizational Performance Review

A review of a couple of key publications in the field might provide an initial data point that could motivate further inquiry and reflection within the field. In turn, various sources of evidence could be used to produce actionable recommendations that will strengthen the field's contributions and credibility.

Performance Measurement and Evaluation

Predictably improving performance depends not only on setting performance goals, and certainly not only on implementing solutions, but also on continuously tracking progress toward desired goals and taking corrective actions as required. This is why we set goals in the first place, to set a direction and track our course to ensure we are still heading in the right direction. This is essentially the role of performance evaluation and measurement.

In one study, performance improvement professionals agreed that identifying or verifying performance goals and objectives—at various organizational levels—is the basis for what they do and how they do it, even if they do not incorporate this into their work as much as they know they should (Guerra, 2003).

What appears to be less obvious, or at least not as popular, is that evaluation and measurement is equally important for improving performance. In the same study, participants attributed less importance to evaluation, particularly with regards to evaluating organizational impact on society (e.g. customers, local community, environment), as suggested by Kaufman (2006).

Evaluation and measurement certainly do not merely occur at the end of implementing an intervention. Rather, these are integral tools for managing and improving performance at every stage of our work. In a classic work, Rummler and Brache (1995) describe the management function at the organizational level as one that:

> *Involves obtaining regular customer feedback, tracking actual performance along the measurement dimensions established in*

the goals, feeding back performance information to relevant subsystems, taking correction action if performance is off target, and resetting goals so that the organization is continually adapting to external and internal reality.(p. 21).

The authors use similar descriptions for the management function at the process level, highlighting the central role of measurement in performance improvement.

The Role of Measurement in Making Decisions.

According to Webster's dictionary, measurement is the estimation of an exact standard. Essentially we use measurement instruments (e.g observation protocols, extant data review protocols; questionnaires, etc.) to compare the object of measurement (e.g. a process, an intervention, a project, etc.) to some pre-specified standard (e.g. goals and criteria).

Measurements give us data that we can then turn into intelligence, and in turn we use intelligence to make sound decisions about what to improve and how.

While fundamental to sensible decision-making, the most neglected aspect of decision making in the literature is intelligence gathering (Eisenhardt, 1998; Nutt, 2006).

Decision making begins when stakeholders see a triggering trend (e.g. declining revenues or sales) or event (e.g. a threat to unionize) as significant, prompting steps to obtain intelligence (Nutt, 2006).

In the performance improvement field, we would say that this would trigger at needs assessment and analysis, where we can measure gaps in results and establish causal factors that would then give us insight as to what solutions are likely to improve performance.

This notion is in fact supported by researchers outside of the performance improvement field, who suggest that signals should be decoded as performance gaps (Pounds, 1969, Nutt, 1979, Cowan, 1986), and that the gap will be considered significant when an important performance indicator, such as market share or revenue, falls below preset criteria and conversely, the signal would be ignored if performance equals or exceeds the expected performance criteria.

When a performance gap is detected, it also reveals the magnitude of the concern to be overcome (Cowan, 1990), this magnitude can be one major consideration in prioritizing performance problems for resolution. Decision making is then undertaken to find ways to deal with closing the performance gap, and reduce or eliminate the concern.

The Role of Measurement in Performance Improvement Processes.

We typically conduct measurement in the context of needs assessment, causal analysis, evaluation, and research.

Performance measurement in the context of needs assessment, allows us to determine the gaps between current and desired performance goals, and in the context of *summative evaluation*, it enables us to determine whether we have reduced or eliminated these gaps through the performance solutions that were implemented.

Further, overarching processes like strategic planning and management entail these aforementioned processes, and thus logically, also entail measurement to a great extent.

Finally, if we clarify the purpose of evaluation, we will also appreciate that evaluation can and should occur at every stage of performance improvement (*formative evaluation*), and again by extension, measurement is at the heart of all that we do.

The Value of Evaluation.

While some rightly say that the fundamental purpose of evaluation is the determination of worth or merit of a program or solution (Scriven,

1967), the ultimate purpose, and value, of determining this worth is in making data-driven decisions that lead to improved performance (Guerra-López, 2007).

The notion that evaluation's most important purpose *is not to prove, but to improve* is an idea originally put forward by Egon Guba decades ago (Stufflebeamx& Shinkfield (2007). Kaufman has similarly proposed that evaluation data should be used to fix rather than to blame (Kaufman & Thomas, 1980).

Along these lines, evaluation is simple:

- It compares accomplished results with planned/expected results;
- It can be used to find driver and barriers to expected performance; and
- Should produce actionable recommendations for improving processes, programs and solutions so that expected performance is achieved and/or maintained (Guerra-López, 2008).

Not only can ongoing measurement and evaluation help us track, manage and improve performance, but also it is through performance evaluation and measurement that we are able to prove the worth of our contributions to our clients and to the field.

Beyond our ability to sell and implement 'solutions', the worth of our contributions are evidenced by the measurable results and ultimate impact that we document.

Purpose of the Study

On this view, measurement and evaluation are at the heart of performance improvement.

This study was intended as a preliminary step toward understanding their current role within the field of performance improvement.

While there are a variety of indicators that could have been chosen, this study focused on the professional literature as an indication of the importance attributed to evaluation and measurement.

For the purposes of this study, the Performance Improvement Journal (PIJ) and Performance Improvement Quarterly (PIQ), as two of the most representative journals of performance improvement interests and foci in terms of the readership and contributing authors, were used as sources.

However, the reader is cautioned to keep in mind that these two professional journals alone do not fully represent the entire field, or its practitioners.

Future studies that look at other relevant journals in performance improvement are warranted to confirm the findings of this study.

The key questions this study sought to answer were:

1. *To what extent do PIJ and PIQ reflect an emphasis on performance measurement?*

2. *Is there a difference in the proportion of articles focused on evaluation and performance measurement between the applied journal (PIJ) and the research journal (PIQ)?*

3. *What is the most prevalent format for these articles?*

4. *Of those articles that emphasize performance measurement, to what extent do they focus on organizational performance vs. measurement of specific solutions?*

METHODS

This study utilized content analysis as a means for answering the research questions.

As a research method, content analysis has seen increased sophistication and utilization in organizational studies (Duriau, Reger, & Pfarrer, 2007).

Examples of a similar trend in the field of Performance Improvement are exhibited in Klein's (2002) study of empirical research in the field and the Marker, Huglin, and Johnsen (2006) replication of it.

According to Duriau et al (2007), content analysis recognizes the relationship between language and attention; this relationship may be examined via various methods, however content analyses generally involve the following processes: (1) data collection, (2) coding, (3) analysis of content, and (4) interpretation of results. This study was conducted according to this framework and included an initial phase of content selection.

Content Selection and Data Review

This study reviewed articles from the Performance Improvement Journal and Performance Improvement Quarterly published over a ten-year period (1997 -2006) and, in an effort to avoid over-reliance upon personal opinion, it excluded editorials, interviews, and guest essays from analysis. This resulted in the inclusion of 792 articles in the review: 545 from PIJ, and 247 from PIQ.

Once articles were identified, the data were collected, coded, and analyzed concurrently. These processes occurred in two phases. During the first phase, the article abstracts (or in the case of PIJ articles, executive summaries) were reviewed in an effort to answer the research questions.

When the review of an abstract yielded an inconclusive response to any of the research questions, the article was set aside for a second phase of more in-depth data review of the entire article.

Of these, only a handful of articles (n=3) were deemed difficult to classify and were set aside for a final and third phase of review approximately a week later as a check for consistency.

All of the data review was performed by the same individual; however a second individual was consulted during the second and third phases of review in order to come to a consensus about final coding. All phases of data analysis were conducted using operational definitions that were carefully discussed and defined by both researchers as follows:

Attention was defined at two levels: *emphasis* and *focus*.

In this case, a performance measurement *emphasis* was indicated by a devotion to the topic of measurement in the article's: (a) abstract (or executive summary), (b) statement of intent, or (c) titles of the headings, tables, figures, or charts included within it. Most of this study was concerned with the emphasis-level of attention.

The issue of *focus* was only relevant for research question #4 (the extent to which authors addressed organizational performance vs. measurement of specific interventions).

Focus was judged based upon which level of evaluation the article advocated; when an article discussed measurement both in support of a particular intervention and organizational performance, this distinction was made based on which area had significantly more content dedicated to it. For example, an article that provided step-by-step guidelines on how to evaluate an electronic performance support system (EPSS) and provided general comments of a few sentences about the importance of measuring organizational performance would be considered to have a measurement emphasis and a focus on intervention measurement).

Performance measurement was defined as measurement activities relating to the needs assessment, analysis, strategic planning, evaluation, or performance tracking contexts.

More particularly, *needs assessment* is a methodology associated with measuring gaps in results; *analysis* is a methodology associated with breaking down elements of performance to identify causal factors; strategic planning includes establishing and tracking strategic measures for the purposes of long-term planning, management, and evaluation; *evaluation* includes determining the effectiveness of organizations as well as specific activities, interventions or anything else meant to contribute to performance; *performance tracking* involves specific measurement and tracking techniques such as development and use of indicators and scorecards.

Format was concerned with the general nature of an article.

Those articles that were *model-oriented* described a particular model for evaluation or measurement; *persuasive* articles advocated *the* use of evaluation or measurement in general or more specifically, the benefits of using a particular model over another; *methodological* articles illustrated *how* to apply evaluation or measurement methodology (e.g. case studies or job aids). When an article exhibited the characteristics of more than one format, this distinction was made based on which area had significantly more content dedicated to it.

FINDINGS

For the sake of clarity, findings have been arranged by each research question as shown below.

1. To what extent do PIJ and PIQ reflect an emphasis on performance measurement?

This research question is comprised of several sub-questions:

Overall emphasis.

PIJ published 188 out 545 (34%) articles related to some aspect of performance measurement, inclusive of the evaluation articles. PIQ published 186 out of 247 (75%) articles related to some aspect of performance measurement, inclusive of evaluation articles.

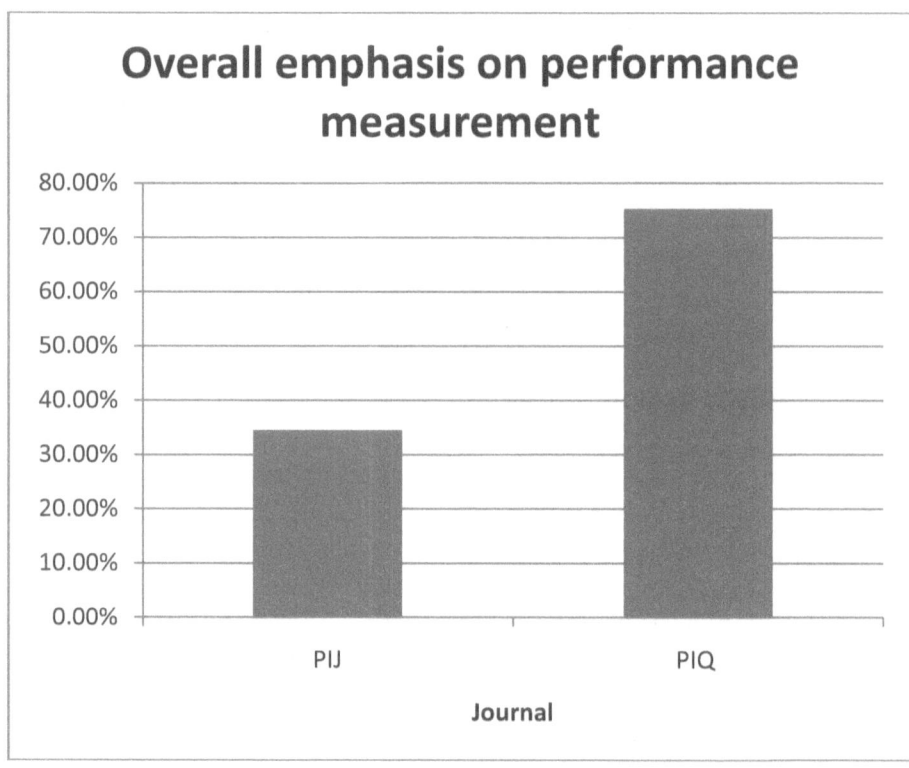

Needs assessment emphasis. PIJ published 15 (2.75%) out 545 articles in the topic of needs assessment, while PIQ published 4 (1.62%) out of 247.

Analysis emphasis. PIJ published 57 (10.46%) out of 545 articles on causal analysis, while PIQ published 59 (23.89%) out of 247.

Strategic planning emphasis. PIJ published 17 (3.12%) out of 545 articles on strategic planning, while PIQ published 7 (2.83%) out of 247.

Evaluation emphasis. PIJ published 57 evaluation articles out a total of 545 articles published, or roughly 10% of publications dealt with evaluation. Meanwhile, PIQ published 108 evaluation articles out of 247 (44%).

Performance tracking emphasis. PIJ published 42 (7.71%) out of 545 articles on performance tracking, while PIQ published 8 (3.24%) out of 247.

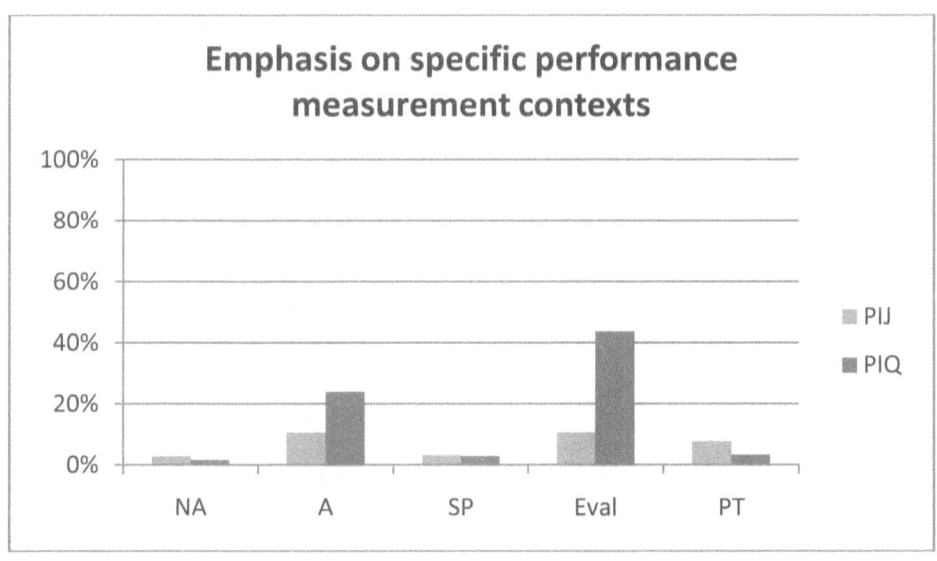

2. **Is there a difference in the proportion of articles focused on evaluation and performance measurement between the applied journal (PIJ) and the research journal (PIQ)?**

Yes, there is an obvious difference with PIQ publishing four times as many evaluation articles than PIJ. In terms of performance measurement in general, PIQ has published over twice as many articles as PIJ. During the content analysis, the changes in editors were also examined as one possible factor for variations in focus upon performance measurement.

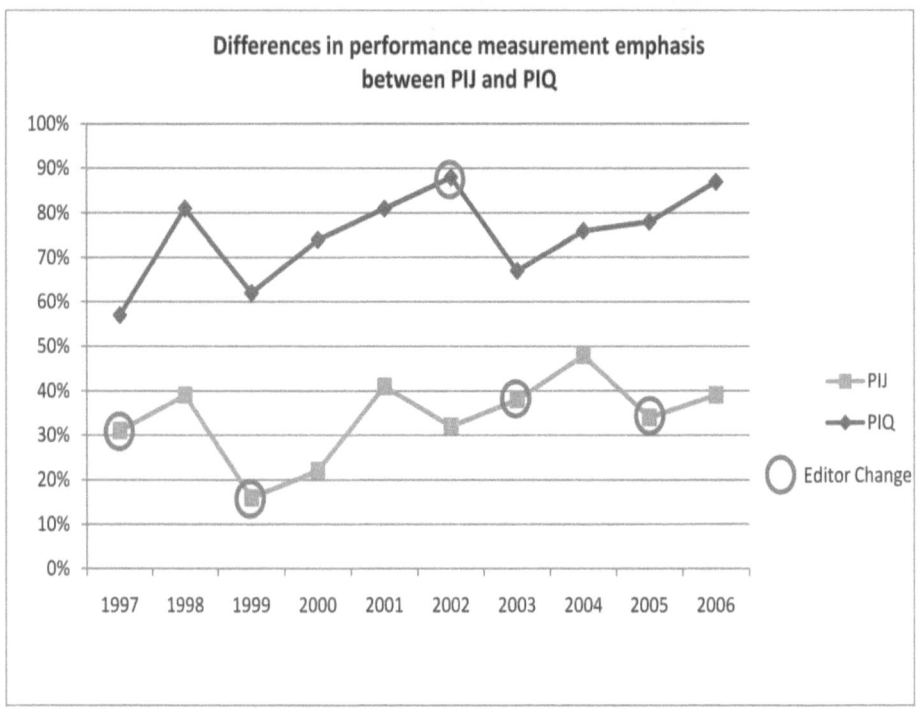

3. What is the most prevalent format for these articles?

Of the 374 emphasizing performance measurement, 246 (66%) of them were methodological, (i.e. they provided guidelines for how to measure a performance, typically through the explanation of a job aid or a case study); the remaining 128 articles were evenly divided between the presentation of a model (n=64) or advocating evaluation in general or a model in particular (n=64)

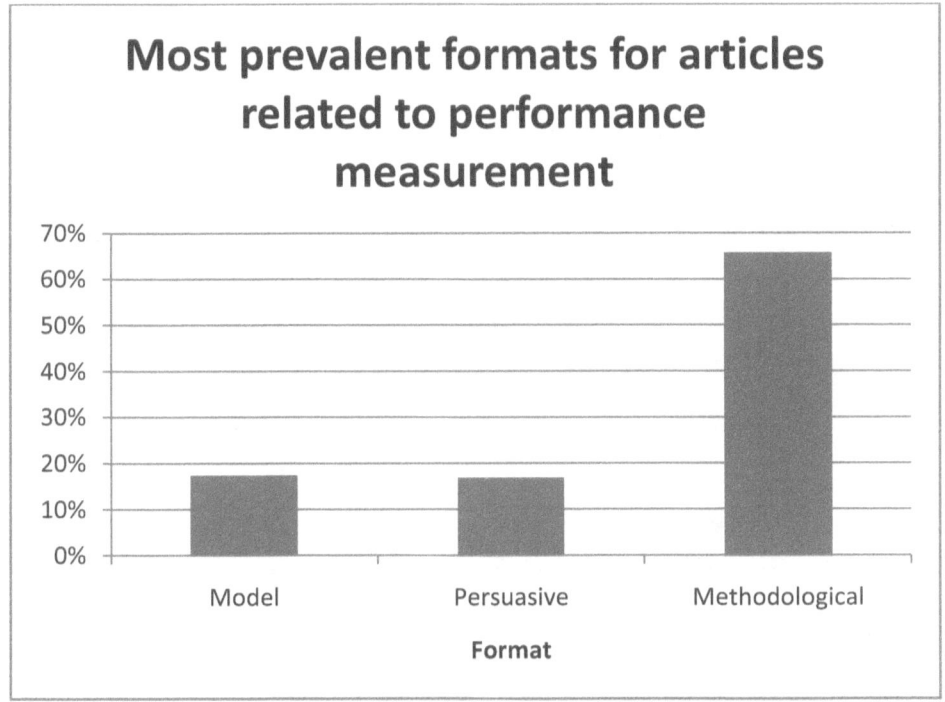

4. **To what extent are performance measurement articles focused on organizational performance measurement vs. measurement of a specific solution?**

In PIJ, 90 out of 188 articles (48%) focused on measurable performance indicators. In PIQ, 47 out of 186 (25%) were focused on measurable performance indicators vs. theoretical, more abstract variables.

In PIJ, 90 (48%) out of 188 focus on performance measurement, while 98 (52%) focus on the actual solution. PIQ published 47 (25%) out 186 that focused on performance measurement, and 139 (75%) related to measurement of a specific solution.

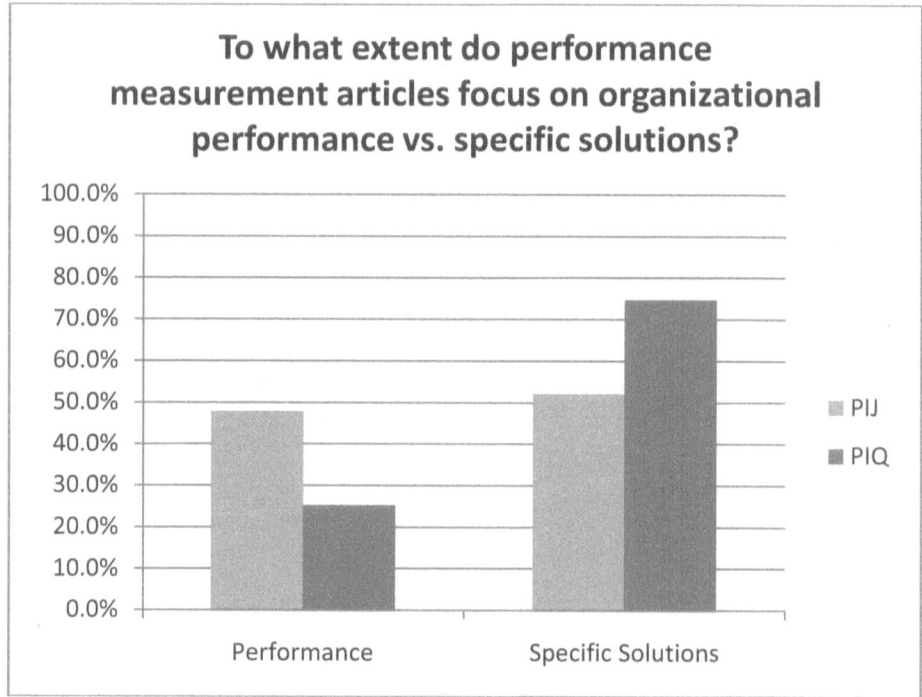

DISCUSSION

Considering that formative evaluation can play an important role in every performance management activity, finding only one tenth of PIJ publications with an evaluation component seems harshly disproportionate. While one would not necessarily expect evaluation to be the central focus of every PIJ article, one might expect to see some evaluative component in much more than one tenth.

PIQ significantly surpassed PIJ in this area with close to half its articles reflecting an evaluation focus. Indeed evaluation research would naturally be thought to belong in this research journal, yet not all evaluations fall into this category. As the practitioner publication, one could have still expected to see more evaluation articles in PIJ that were perhaps less formal or structured.

The performance measurement article ratios are expectedly higher, since for the purposes of this study, we considered evaluation as a particular category within performance measurement.

One third of PIJ articles were indeed directed at some aspect of performance measurement suggesting a relative awareness of measurement as an important component of performance management and improvement. The PIQ results illustrated an outstanding two thirds of articles focused on measurement.

This certainly does not come as a surprise, since measurement is a central mechanism in research, and PIQ is a key research journal of the field. If we follow this argument, however, it is interesting then, that the findings did not reveal an even higher ratio. One issue that could account for not finding more PIQ articles that were focused on measurement is the fact that the remaining articles were for the most part qualitative case studies. Both qualitative and quantitative studies have the potential to be equally strong and valid, provided they are conducted rigorously and appropriately. However, while quantitative research is generally centered on measurement, qualitative research is generally centered on descriptions.

Since PIJ tends to be considered the "practitioner" journal in the performance improvement field while PIQ is considered the "research" journal, it may not be too surprising to find a significantly higher "talk" of measurement in the research journal. Yet, measurement does not exclusively take place in a traditional research context.

Measurement is at the core of what all serious performance consultants do (Rummler, 2004). If practitioners are indeed doing it, why are their evaluation and measurement practices not captured in their articles, even at an anecdotal level? We should at least expect to see publications about measurement practices as much as we see publications on interventions. After all, one of the characteristics that stands us apart from other types of consultants, and indeed typical sales people, should be the emphasis on actual improvement…not just selling and/or implementation trends.

The primary differences between PIQ and PIJ should not necessarily be the topics themselves, but rather whether the topics are being discussed in the context of a research study, or in a more applied setting. Evaluation and measurement can/should be done in both contexts.

The fact that almost two thirds of measurement articles are focused on providing the readership with "how-to's" is a promising indication that there are indeed performance improvement professionals advocating and disseminating measurement practices. While we cannot assume that these are "best practices", it is certainly a step in the right direction and could be used as the first—awareness-raising—step in encouraging the readership to follow up on how to integrate measurement into their practices.

Essentially, PIJ has a half and half split between articles focused on measuring organizational performance vs. measuring the performance of a specific solution. The focus on measurement in PIQ on the other hand is primarily focused upon testing the performance or effectiveness of a specific solution. This may not be an unexpected result as PIQ publishes research articles, and by their very nature require articulation of specific variables. Most of the time, these independent variables, that is, the interventions, are tested for specific impact, rather than focusing on general organizational performance variables.

In regard to the effect of editorial change on performance measurement emphasis, two of three decreases in PIJ occur in years where a change in editor occurs. On the other hand, PIQ has fewer editorial changes. Furthermore, given PIQ's review peer review process for selecting articles, one would suspect that journal's emphasis to be less susceptible to editor change. Additionally, if we consider special issues and guest editing, editor change does not paint the whole picture.

CONCLUSION AND RECOMMENDATIONS

This study was intended as a preliminary step in uncovering the perceived value of performance evaluation and measurement for performance improvement professionals.

The data, in particular that related to our practitioner journal, PIJ, reveals that our attention to evaluation and measurement is not at a level that supports our claims to add measurable value to our clients. Its publications show relatively low attention to performance measurement, and even lower attention to evaluation in particular.

If evaluation and measurement are not central mechanisms of our performance improvement projects, how are we supporting our claims to add measurable value to our clients?

This study also included a preliminary consideration of the impact of editor change on the measurement and evaluation focus within the journals. It is likely that editors' opinions and values influence the selection of articles, but these data do not conclusively support a causal linkage between the two. Rather, the data suggest that more study in this area is warranted, especially as it may guide editorial selection practices.

It would behoove performance improvement professionals to not only talk about evaluation and measurement but to actually integrate it into their work. If we want future analysis of our practices and solutions to support our professional and ethical credibility, we must have the data to support our claims.

Performance data tracked through evaluation and measurement should be readily available at any given time for any of our projects, even if "the client will not pay for it."

Tracking the effectiveness of our own work should be ingrained into what we do anyway. For example, in the context of managing our projects, we must track what was planned against what was accomplished.

It does not necessarily have to be considered a separate activity and deliverable for the client, though they certainly will benefit in many ways, including justifying decisions with relevant, reliable, valid, and not least of all well documented data.

We can build time into each of our other project phases and call it whatever we want, not necessarily "evaluation" or "performance tracking." What is important is that it gets done and used to improve every phase of our projects, including the final products and contributions we deliver.

It is worth reiterating that the literature is but one indicator of the practices of a field. As with the interpretation of any other data set, it must be verified and understood in the context of other relevant, reliable, and valid indicators before confidently making interpretations and drawing conclusions (Guerra-Lopez, 2008).

Readers are also cautioned to keep in mind that while operational definitions and a review protocol were used to categorize and count the articles, the findings are dependent upon subjective judgments and the results could have been different had another observer used the same tools to make the observations. Replication of this study would be advisable in order to confirm findings.

Finally, it must be noted that PIJ and PIQ do not necessarily represent the values, intentions and practices of every performance improvement professional. Rather the articles reviewed for this study are more directly a reflection of the authors, editors, and editing review boards that accept manuscripts for publication.

Bibliography

Andrasik, F. (1979). Organizational Behavior Modification in business settings: A methodological and content review. Journal of Organizational Behavior Management, 2, 85-102.

Balcazar, F. E., Shupert, M. K., Daniels, A. C., Mawhinney, T. C., & Hopkins, B. L., (1989). An objective review and analysis of ten years of publication in the *Journal of Organizational Behavior Management, 2*, 7-37.

Clark, R. & Estes, F. (2000). A proposal for the collaborative development of authentic performance technology. *Performance Improvement*, 39(4). 48-53.

Cowan DA. (1990). Developing a classification structure of organizational problems: an empirical investigation. Academy of Management Journal; 33(2):366–390.

Culig, K. M., Dickinson, A. M., McGee, H. M., & Austin, J. (2005). An objective comparison of applied behavior analysis and organizational behavior management research. *Journal of Organizational Behavior Management, 25(1)*, 35-72.

Duriau, V. J., Reger, R. K., & Pfarrer, M. D. (2007). A content analysis of the content analysis literature in organizational studies: Research themes, data sources, and methodological refinements. Organizational Research Methods, 10(1), 5-34.

Eisenhardt K. (1998). Decision making and all that jazz. In: Papadakis V, Barwise P, editors. Strategic decisions. Boston, MA: Kluwer.

Frederiksen, L. W., & Johnson, R. P. (1981). Organizational Behavior Management. Progress in *Behavior Modification, 12*, 67-118.

Frederiksen, L. W., & Lovett, S. B. (1980). Inside Organizational Behavior Management: Perspectives on an emerging field. *Journal of Organizational Behavior Management*, 2, 193-203.

Guerra, I (2003). Key Competencies Required of Performance Improvement Professionals. Performance Improvement Quarterly. 16 (1).

Guerra-López, I. (2007). Evaluating Impact: Evaluation and Continual Improvement for Performance *Improvement Practitioners*. Human Resource Development Press.

Guerra-López, I. (2008) Performance Evaluation: Proven Approaches for Improving Program and Organizational Performance. Jossey Bass.

Kaufman, R. (2006). *Change, Choices, and Consequences: A Guide to Mega Thinking and Planning*. Amherst, MA. HRD Press Inc.

Kaufman, R. & Clark, R. (1999). Re-establishing performance improvement as a legitimate area of inquiry, activity, and contribution: Rules of the road. *Performance Improvement*. 38(9). 13-18.

Kaufman, R. & Thomas, S. (1980). *Evaluation without Fear*. New York, NY.New Points/Franklin Watts

Klein, J. D. (2002). Empirical research on performance improvement. *Performance Improvement Quarterly*, 15(1), 99-110.

Hopkins, B. L., & Sears, J. (1982). Managing behavior for productivity. In L. W. Frederiksen (Ed.), *Handbook of Organizational Behavior Management* (pp. 393-425). New York: John Wiley.

Maker, A. Hughlin, A. Johnsen, L. (2006). Empirical Research on Performance Improvement: An Update. *Performance Improvement Quarterly*. 19(4). 7-22.

Nutt P. (1979). Calling out and calling off the dogs: managerial diagnoses in organizations. Academy of Management Review ;4(2):203–14.

Nutt, P. (2006). Intelligence gathering for decision making. Omega N. 25 (604-622).

O'Hara, K., Johnson, C. M., & Beehr, T. A. (1985). Organizational Behavior Management in the private sector: A review of empirical research and recommendations for further investigation. *Academy of Management Review, 10*, 848-864.

Pounds W. (1969). The process of problem finding. Industrial Management Review 1969; 1–19.

Rummler, G. A. (2004). *Serious performance consulting: According to Rummler*. Silver Spring, MD: International Society for Performance Improvement.

Rummler and Brache (1995). *Improving performance: How to manage the white space on the organization chart*. 2nd Ed. Jossey-Bass: San Francisco

Scriven, M. (1967). The methodology of evaluation.. In R. Tyler, R. Gagne, and M. Scriven (eds). *Perspectives on curriculum evaluation*. New York: McGraw Hills.

Stufflebeam, D. & Shinkfield, A. (2007). *Evaluation theory, models, and applications*. San Francisco: John Wiley & Sons.

Ingrid Guerra-López, PhD

Dr. Guerra-López is an Associate Professor at Wayne State University, Associate Research Professor at the Sonora Institute of Technology in Mexico, and Director of the Institute for Learning and Performance Improvement.

Dr. Guerra-Lopez's research, teaching, and consultancy focuses on improving performance and management decision-making through the design, development, and use of performance measurement, tracking and management systems.

She is also Principal of Intelligence Gathering Systems.

Ingrid has written five books on performance evaluation and assessment, as well as published nearly thirty articles and ten book chapters on performance improvement, assessment, and evaluation.

Hillary N. Leigh

Hillary Leigh is a doctoral student in Wayne State University's Instructional Technology program.

Her research interests include evidence-based intervention selection and justification for the field of Performance Improvement.

Her dissertation topic relates to practitioners' usage of scientific and artistic evidential sources when selecting an intervention.

On the practical side, she has consulted with healthcare, educational, and retail organizations to select, develop, implement, and evaluate a variety of instructional and non-instructional interventions.

Social Responsibility of a Profession: An Analysis of Faculty Perception of Social Responsibility Factors and Integration into Graduate Programs of Educational Technology

By

Stephanie L. Moore, PhD.

Introduction

Asking the question of ethics is one of asking "What is good?" At the end of the day, when our work as educational technologists is done, when may we say to ourselves, "Well done"? When the technology is adopted? Not hardly.

When the system has changed and adapted? That depends. Ely (1976) states that "Neither stability nor change have any intrinsic value.

"The worth of stability is in the goodness it preserves, while the worth of change is in the goodness it brings about" (p. 151).

This maxim, of sorts, provides an excellent crux for ethical stance educational technologists can adopt.

Asking the question of ethics is one focused on caring about our craft and its impact on society.

Caring about one's work is the internal aspect that creates "good" in what one produces.

When dealing with the process of asking members of a system to modify their thinking and processes or actions, this introduces several notions of ethical obligations aimed at doing "good" work that reflects true caring. In fact, we can begin to operationalize "good" in the practice of educational technologists.

The purpose of this study is to provide an empirically-driven definition of "good" in our profession and examine the presence of it in graduate programs. If the technologists do not care about their work, then it is highly unlikely that others using technology will perceive quality in the product.

Research shows that ethics in action do indeed increase quality and raise the bar for performance, both enhancing financial bottom lines and social bottom lines at the same time (Dobni, Ritchie & Zerbe, 2000; Kaufman, 1997, 2000, 2006; Verschoor, 1998). Ethics are more than philosophical pondering of some abstract concept; they are a means of defining exemplary performance standards that can be expected of members in a profession (Dean, 1993).

Driving Forces for Ethics in Professional Courses

Davis (1999) describes what he calls an "ethics boom" in higher education during the past 30 years, where one-by-one professions have been faced with a requirement to integrate ethics explicitly in university programs.

Technological advances, national scandals and poor performance have driven other professions to build ethics into college curriculum as a means of ensuring good decision making and defining standards for members of a profession.

The medical profession was among the first to identify a gap of ethics integration into college curricula (Davis, 1999, p. 4). Advances in medical technology confronted practicing physicians with difficult decisions, as they had to choose what deserved budget allocations.

While one machine could save the lives of a few dozen patients, the equipment was so expensive that the same amount of money could also build an outpatient clinic that would serve many more people but not save lives. To address these issues directly in practice and train the next generation of physicians for these decisions, the medical community turned to philosophy departments to help them build courses on practical ethics.

For the legal profession, it became a public point of embarrassment that many of President Richard Nixon's team members involved in the criminal activities of Watergate were lawyers.

The legal profession faced public demand for ethics in their practices and training. Some states started requiring ethics courses as a condition for admission to the bar, other states quickly followed suit, and law schools suddenly found themselves facing social and legal requirements to teach ethics to their students (Davis, 1999, p. 6).

Davis states that the law profession was a little different from the medical field because the law literature had developed ethical concepts; however, they were still caught in the same problem of definition: how do codes of ethics translate into professional responsibilities and performance standards?

As the profession grappled with distinctions, courses on ethics in major national programs became the norm and remain a staple requirement today.

Soon, a cluster of engineering and science disciplines followed suit. Spiro Agnew, Richard Nixon's Vice President, had resigned office because of a bribing scandal involving numbers of engineers during his office term in Maryland.

Civil engineers seeking state contracts had bribed Agnew to obtain those contracts.

When news spread, *"Engineers all over the country were appalled that so many engineers could be involved in such a flagrant violation of professional ethics"* (Davis, 1999, p. 7).

Davis explains that engineers were already uneasy because their field had been involved in fake testing on Goodrich's A-7D airbrakes, the Ford Pinto's exploding gas tank, and the DC-10s misdesigned cargo door.

From this period forward, many disciplines started following the same pattern – business, accounting, nursing, journalism, financial analysis, public administration, dentistry and others. The trend is clearly toward integration of ethics into college curricula, specialized to each discipline (Davis, 1999, p. 7).

Frankel (1989) explains that in the recent past, discussions on professional ethics occurred only within the professions themselves. Historically, professions have maintained a negotiating process with society, keeping a tension between the autonomy of a profession to define and regulate itself and the public's demand for accountability.

He states, *"Society's granting of power and privilege to the professions is premised on their willingness and ability to contribute to social well-being and to conduct their affairs in a manner consistent with broader social values"* (p. 110).

While the educational technology and human performance technology may not have had front-page scandals that dominated media outlets for an extended time, there are publicized reports (*A Nation at Risk*, 1983, U.S. Department of Education – National Commission on Excellence in Education), recent policy (such as *No Child Left Behind*), and vocal critics (Cuban, 1986, 2001; Healy, 1990, 1999) that all underscore a perceived failure of technology advocates to contribute anything worthwhile to educational or learning systems.

Early Educational Technology Literature: Ethics Recognized by Founding Authors

Attention to the requirement for educational technology to contribute to the "worthwhileness" of learning and performance environments (Davies, 1996, p.16) is a long-time topic in the literature, appearing in foundational pieces for the field. Authors such as Finn (1996a and 1996b), Davies (1996) and Kaufman (1969, 1996, 2000a, 2006) explicitly raised awareness on ethics for the profession, calling for a professional code of ethics, reflection on the ethical nature of educational technology, and assessment procedures that ensured the profession ultimately contributed to society (Ely & Plomp, 1996).

Founding literature did recognize that technology was significantly altering the landscape of education, and those changes called for an enhanced sense of moral obligation.

According to Davies (1996, original published in 1978), while technology and creativity expanded the range of choices available to educators, they also "made it more difficult to foresee the full consequences of the choices made and the actions taken" (p. 15). He states:

> *"Technology, contrary to popular belief, is not necessarily confined to the means (sic) by which educators realize their ends. Technology also raises anew questions about the nature of the ends themselves. It forces us to reflect on the morality of what we are about, by its very insistence on defensible choices".* (sic, p. 15)

Finn states that technology is not a collection of gadgets, hardware and instrumentation but is instead "a way of thinking about certain classes of problems and their solutions," (p. 48) underscoring the notion of a results orientation.

For Davies and Finn both, this results orientation maintains the question of "What is desirable, and why is it desirable?" – a question they believe best answered through continual contemplation (Davies, 1996, p.17), or philosophical examination (Finn, 1996a, p. 49).

What Finn addressed through philosophical examination, Kaufman addressed through an assessment approach.

Asserting the importance of a results orientation, Kaufman (1996, reprinted from 1977; 2000a, 2006) provides a practical way of discussing results (or ends) and societal benefit by framing this discussion in terms of assessment.

Kaufman outlines the explicit relationship between what educational technologists do (or what education planners in general do) and the ultimate impact of such work on society:

> *"The simple truth is that what the schools do and what the schools accomplish is of concern to those who depend upon the schools, those who pay the bills and those who pass the legislation.*
> *We are not in a vacuum, and our results are seen and judged by those outside of the schools – those who are external to it.*
> *...*
>
> *This external referent should be the starting place for functional and useful educational planning, design, implementation, and evaluations – if education does not allow learners to live better and contribute better, it probably is not worth doing, and will probably ending up being attacked and decimated by taxpayers and legislators." (1996, p. 112)*

Thus, according to Kaufman, the practice of educational technology should first begin by determining and justifying what the ultimate *desirable* impact of our actions are on society and using that as a guide for the design process. Kaufman has developed this over the years into a full framework for assessment that he calls "Mega" (2000b, 2006), which may very well prove to be an *ethical* framework for the field given its focus on social impact.

As the field took shape, Davies (1996), Finn (1996b), Kaufman (1969, 1996) and others made it explicit that this question of social responsibility – of the profession's ultimate impact on society and the results of our actions – was something that the field must answer to if it was to be a viable, respectable profession.

Ethics as Performance Standards for Professionals

Peter Dean (1999), an author on ethics research in Human Performance Technology (HPT) literature, argues that the study of ethics is critical in two ways.

First, ethical codes are behavioral guidelines for professionals (p. 703). They "reflect and support the ethical values of the organization, clarify expectations, and recognize specific ethical issues" (p. 703).

Codes of ethics may not eliminate unethical behavior, but they do establish clear expectations for performance, and, as Dean points out, *"Isn't establishing clear performance expectations one of the first things we recommend for achieving exemplary performance?"* (Dean, 1993, p. 5).

Second, HPT professionals are in a unique position to help clients and organizations achieve their goals ethically and guide them toward ethical actions from the start (Dean, 1993, p. 5). By the very nature of HPT work, professionals can help organizations and clients *"recognize that the ethical climate in an organization is one of the environmental factors that impact performance and productivity"* (p. 5).

Research in business ethics demonstrates that practitioners who have been trained in ethics do perform better on the job and tend to have higher rates of productivity (Delaney & Sockell, 1992). Dobne,

Ritchie, and Zerbe (2000) examined the relationship between organizational value systems and employee productivity. They found that in systems where values were clear and articulated, employees had a higher level of commitment and higher positive affect about their jobs.

Research in business ethics also showed that the way ethics were implemented or modeled made a difference in the ethical behavior of employees. It is not enough to have a code of ethics, and it doesn't work to punish employees when they fail to adhere to the code (Dean, 1993).

Findings in Weaver (1999) suggest that leadership in and modeling of ethical behavior, versus simply demanding it, has a greater impact on the ethical behavior of employees. For application in educational programs, these findings would suggest that students will exhibit more ethical behavior if faculty model that same behavior.

Higher Education: The Place for Ethics Training

Cleary, as ethical issues have been identified and as ethical decision-making has been studied, professions have turned to higher education programs as the place for "improving" the ethics of future generations of professionals.

The role of ethics in professional education has long been noted as essential, even stemming back to Socrates, who said that no craft or profession should seek its own advantage but should benefit those who are subject to it (Baumgarten, 1982).

Baumgarten argues that, through university teaching, *"we express our conviction that thoughtful inquiry ennobles a human life and contributes to human excellence ... there is special reason to value a profession that is solely committed to enlarging the power and influence of reason discourse and imaginative questioning"* (p. 294).

Professional ethics pertain to members of a particular profession, and entrance into that profession is gained only through some form of advanced study. Thus, the place of advanced study is the very place where professional ethics must be learned – either as a formal part of the education or in some less formal way in job settings.

Furthermore, universities are viewed as the place where training in ethics should take place prior to graduates entering the workforce. Procario-Foley and Bean (2002) argue that organizations require the recruitment of "graduates who have world-class ethics to accompany their world-class knowledge" (p. 105).

Indeed, this reflects recent findings that over 60% of organizations rank ethics as one of the top priorities they look for in candidates for open positions (Hammond, 1992; cited in Dean, 1993). Caron (1999) urges that higher education institutions are in a unique position to train graduates on how to address social concerns without having to adopt any particular political agenda, and can do so with intellectual rigor and professional capacity.

Finally, research indicates that the ethical philosophies and values (and resulting behaviors) of management or leaders influences ethical choices and behaviors of employees and other organizational members (Brenner & Molander, 1977; Ford & Richardson, 1994; Petrick & Quinn, 1997; Procario-Foley & Bean, 2002).

We can translate the language of business to education programs, where faculty in leadership and mentorship positions model philosophies and behaviors that may influence students in their programs.

O'Connell (1998) states, *"Our task in universities is not only to teach ethics and values for the marketplace but to model these values ourselves as we fulfill our own moral responsibility as educators in the universities where our students begin the business ethics journey in the first place"* (p. 1620).

Procario-Foley and Bean (2002) explain that students are keenly aware of the ethical behaviors their course instructors demonstrate. They state, *"Teaching faculty are exemplars for students and it is essential that they reflect and personify the values of the institution"* (p. 112).

Matchett (2008) asserts that ethics *are* modeled to students by faculty on a daily basis in universities: *"All colleges teach ethics across their undergraduate curriculums, yet relatively few institutions do this deliberately"* (p. 25).

Matchett explains that this may not necessarily be bad for students, but it bears examination since research has shown the number of formal years of education to be a powerful predictor of moral judgment beyond any other variable (McNeel, 1994; Nucci & Pascarella, 1987: Rest 1994).

However, failure to be deliberate in the approach to teaching ethics may have a number of unintended consequences (Matchett, 2008, p. 25). Thus, college programs are the place where professional ethics are both taught and modeled, on a practical day-to-day basis.

Purpose of the Study

The purpose of this study was to determine the degree to which the ethic of social responsibility, as defined as adding measurable value to our shared society, is present in United States and Canadian graduate programs in educational technology and examine possible factors and barriers that explain the absence or presence.

This section presents the findings to three research questions aimed at answering the overall question of whether social responsibility is on the professional development radar in educational technology.

As Kaufman (1999, 2006) and Brethower (2005) continue to assert, if you are not adding value to society, you are likely subtracting value from society. Thus, it is not safe to assume that the educational technology profession is necessarily benefiting society without a means for determining the accuracy of that assumption. As a profession, the field of educational technology has a social contract with society in general to ensure that it benefits society and does not harm individuals or the social system in its practices. The results presented in this chapter provide a first glance at where the profession of educational technology as practiced in North America is in relation to that possible ethics boom in professional studies.

Research Design

The design of this study was nonexperimental, involving descriptive statistics and correlational designs. Gall, Gall, and Borg (2003) state that questionnaires and interviews are commonly used to collect data about "phenomena that are not directly observable: inner experience, opinions, values, interests, and the like" (p. 222).

Because this research design is primarily descriptive on unobservable phenomena such as philosophies and beliefs and will be sent to members covering a wide geographic area, a questionnaire design is appropriate.

An online questionnaire was selected based on ease of administration and elimination of paper and mailing costs (Gall, Gall, & Borg, 2003, p. 230).

Because this questionnaire was delivered to faculty in a technology-related field, it was assumed that participants had email as a means for distributing the questionnaire and access to computers as a means for completing it. The questionnaire was developed on a server with the capability of collecting data electronically and ensuring integrity of the data.

Participants

Correlational statistics were used to answer research question 3. To detect a medium effect size using correlational statistics with α = .05 and β = .30 (power set at .70), Gall, Gall, and Borg (2003) state that the sample size should be 66 (see table on p. 143). The research questions and findings reported here are a subset of the analyses conducted on a broad study (Moore, 2005). For the full study, some more demanding statistics were involved, so the statistical power of analysis demanded a higher sample size.

Sampling

The population of interest in this study was faculty teaching in educational technology graduate programs.

Programs were identified through the listings in the *Educational Media and Technology Yearbook* (2003, 2004). The *Yearbook* categorizes programs by emphasis areas and provides detailed information on the programs, such as what degrees they offer and how many faculty teach (full-time and part-time) for the institution. The publication includes programs in educational technology, educational media, instructional design, information and library sciences and computer information systems.

For the purposes of this study, only those programs with primary emphasis in educational technology/media or instructional design were chosen.

The only programs not included in the study were those for which no faculty contact information was available or faculty were not easily identifiable in the target programs. A total of 694 faculty were sampled for this study. After one initial email invitation and three reminders, 169 participants (25% response rate) completed the survey.

The following tables provide demographic data of the participants most relevant to this discussion:

Table 1
Primary Teaching Emphasis Frequencies and Percentages

Primary Teaching Emphasis	N	Percentage
Instructional Design	35	22%
Multimedia Development	16	9%
Change/Technology Integration	26	15%
Systems Design	2	1%
Human Performance Technology	10	6%
Critical Theory	2	1%
Assessment	0	0%
Evaluation	8	5%
K-12 Media	16	9%
International Studies	0	0%
Research	17	10%
Distance Learning	9	5%
Other	27	16%
Blank	1	1%
Total	169	100%

Table 2
Professional Associations Frequencies and Percentages

Professional Association Membership	N	Percentage
AECT	84	49.7%
ISTE	58	34.3%
ISPI	26	15.4%
AERA	71	42%
APA	8	4.7%
ASTD	19	11.2%
None of the Above	26	15.4%
Blank	1	0.6%

Instrument: Perception of Societal Impact

One existing instrument was used for faculty perceptions of educational technology's impact on society, adapted from Kaufman

(2000): a gap analysis for What Is and What Should Be for the frequency of societal impact.

This instrument consisted of 13 questions, using the same concepts from the elements of the Basic Ideal Vision.

Kaufman (2000, 2006) suggests that the HPT professional community use his instrument as an initial measure for determining how committed they are to delivering positive societal impact as a professional in the field.

The instrument has a Likert-type scale on either side of each question, both ranging from 1-6 (1=Does Not Apply, 2=Rarely, if ever, 3=Not Usually, 4=Sometimes, 5=Frequently, 6=Consistently). On the left side, respondents indicated "What Is," describing what their practices currently looked like. On the right side, respondents indicated "What Should Be," describing what they believed their practices should look like.

A note on how some of the answers were coded and analyzed is in order. Very early faculty responses to the instruments indicated that some were uncomfortable with the items in Kaufman's Elements, such as Murder and Rape.

Even though they could chose "Does Not Apply," their visceral reactions to some items indicated that "Does Not Apply" did not really capture their response. Thus, another option was added of "I prefer not to answer this question." The response was coded as a -1 in the raw data but treated as a 0 for computations.

Additionally, when participants skipped a question entirely (not marking anything), that answer was recorded as a 0. For the purposes of this study, the two responses were treated as a 0 because they were interpreted as participants' desires to simply not think about the item or consider how it does or should apply in their work. Thus, a 0 in the analysis was treated as a further distancing of one's self from social responsibility than a "Does Not Apply." The implications of this decision are discussed later in implications for future research.

Research Question 1: With what frequency do faculty report they should be committed to societal impact (Kaufman, 2000) as a professional in educational technology?

The researcher assessed the frequency with which faculty reported they should be committed to societal impact by reviewing the mean and median ratings for each item. Because this variable was measured on an ordinal scale, the median was determined to be the most appropriate statistic of central tendency (Ary, Jacobs, & Razavieh, 1996, p. 139; Guerra, 2001).

The median is not sensitive to extreme scores, so it is an appropriate index for finding the typical score.

Although the mean is not well suited for ordinal data, means were also estimated an included in this chapter for comparison purposes.

Mode was determined to be an inappropriate measure for central tendency as well because it is very unstable, and a distribution can have more than one mode (Ary, Jacobs, & Razavieh, 1996, p. 136). Use of mode is usually limited just to inspection purposes, not reporting purposes (Ary, Jacobs, & Razavieh, 1996, p. 136).

One item received a median rating of 2 (Never): war and/or riot. Seven items received a median rating of 3 (Rarely): shelter; murder, rape and crimes of violence or destruction; substance abuse; disease; pollution; child abuse; and discrimination. One item received a median rating of 3.5 (between Rarely and Sometimes): Consequences.

Three items received a median rating of 4 (Sometimes): unintended human-caused changes to the environment, starvation and/or malnutrition, and accidents. One item received a median rating of 6 (Always), the highest score on the scale: partner/spouse abuse. See Table 24 for a summary of median and mean scores for each item.

Research Question 2: With what frequency do faculty report they are   committed to elements of societal impact (Kaufman, 2000) as a professional in educational technology?

The researcher assessed the frequency with which faculty reported they are committed to societal impact by reviewing the mean and median ratings for each item.

All items but one received a median score of 2 (Never) or 3 (Rarely). Six items received a median score of 2 (Never): war and/or riot; murder, rape and crimes of violence or destruction; substance abuse; disease; pollution; and discrimination. Six items received a median score of 3 (Rarely): shelter, unintended human-caused changes to the environment, starvation and/or malnutrition, child abuse, accidents, and Consequences. One item received a median score of 5 (Usually): partner/spouse abuse. See Table 24 for a summary of median and mean scores for each item.

Research Question 3: What, if any, is the relationship between how frequently faculty think they should be and are committed to societal impact?

To determine the relationship, the medians and means of "What Is" responses were subtracted from the "What Should Be" responses. Table 24 summarizes those gaps for all items. For every item, the "What Should Be" ratings were higher than the "What Is" ratings.

Using the median as the unit of comparison, the largest gap was 1, which was obtained for nine items.

One item had a median gap of 0.5. Using the mean as the unit of comparison, the largest gap was 0.56, which was obtained for one item: unintended human-caused changes to the environment.

The next largest gap was 0.48, obtained for one item: accidents. The third largest gap was 0.46, obtained for one item: Pollution.

Table 3

Median and Mean Responses of Faculty in Educational Technology

1=Does Not Apply 2=Never   3=Rarely   4=Sometimes
                  5=Usually   6=Always

(0="I prefer not to answer")

Questionnaire Item	What Is			What Should Be			Gap	
	n	M dn	M	n	M dn	M	M dn	M
War and/or riot	154	2	2.47	154	2	2.76	0	0.29
Shelter	154	3	2.79	154	3	3.11	0	0.32
Unintended human-caused changes to the environment, including permanent destruction of the environment and/or rendering it non-renewable	154	3	3.23	154	4	3.79	1	0.56
Murder, rape, or crimes of violence, robbery, or destruction of property	154	2	2.63	154	3	2.91	1	0.28
Substance abuse	154	2	2.61	154	3	2.92	1	0.31
Disease	154	2	2.38	154	3	2.73	1	0.35
Pollution	154	2	2.55	154	3	3.01	1	0.46
Starvation and/or malnutrition	154	3	3.08	154	4	3.39	1	0.31
Child abuse	154	3	2.83	154	3	3.10	0	0.27
Partner/spouse	154	5	4.52	154	6	4.88	1	0.36

abuse								
Accidents, including transportation, home and business/workplace	154	3	2.93	154	4	3.41	1	0.48
Discrimination based on irrelevant variables, including color, race, creed, sex, religion, national origin, age, and location	154	2	2.54	154	3	2.78	1	0.24
Consequences of the Basic Ideal Vision: [76]	154	3	2.42	154	3.5	2.85	0.5	0.43

In addition to gap scores, the relationship between "what is" and "what should be" scores was also explored through statistical means using correlational analysis.

The results of Spearman's Rho are presented in Table 25. For all 13 items, the "what is" and "what should be" scores had high correlations that were significant at the .01 level (2-tailed).

---

[76] Any and all organizations – public and private – will contribute to the achievement and maintenance of this Basic Ideal Vision and will be funded and continued to the extent to which they meet its objectives and the Basic Ideal Vision is accomplished and maintained. People will be responsible for what they use, do, and contribute and thus will not contribute to the reduction of any of the results identified in this Basic Ideal Vision.

Table 4

Spearman's Rho correlations for "What Is" and "What Should Be"

Item	Spearman's Rho
War and/or riot	.946
Shelter	.920
Unintended human-caused changes to the environment, including permanent destruction of the environment and/or rendering it non-renewable	.826
Murder, rape, or crimes of violence, robbery, or destruction of property	.917
Substance abuse	.896
Disease	.929
Pollution	.875
Starvation and/or malnutrition	.876
Child abuse	.933
Partner/spouse abuse	.673
Accidents, including transportation, home and business/workplace	.838
Discrimination based on irrelevant variables, including color, race, creed, sex, religion,	.868

national origin, age, and location

Consequences of the Basic Ideal Vision: **Any and all organizations – public and private – will contribute to the achievement and maintenance of this Basic Ideal Vision and will be funded and continued to the extent to which they meet its objectives and the Basic Ideal Vision is accomplished and maintained. People will be responsible for what they use, do, and contribute and thus will not contribute to the reduction of any of the results identified in this Basic Ideal Vision.**	.911

Discussion

The results were that faculty believed the different elements applied Rarely or Never in all cases but one.

In one case, the element of Spouse/Partner abuse was rated highly both in terms that faculty *should be* committed to this and *are* committed to it.

Comments and feedback from faculty corroborated the notion that most faculty just do not see these elements has having anything to do with what they teach and do in the field of educational technology. Sample comments from faculty that support this interpretation include:

1. *"I care about those issues and do some small things to combat them (and especially not add to them) but I don't see my having much chance or responsibility to change those things in my intructional technology classes. The closest I can come to that is to see that people understand issues like the digital divide and that the deaf/blind are not disenfranchised by bad technological decisions." (sic)*

2. *"I was very confused by the Elements of Social responsibility section of this survey. Although the design of this portion of the instrument was creative, I was often unable to see the connection to instructional technology."*

3. *"The first section on ethics did not relate at all to our educational technology program."*

4. *"That last set of questions...well I just don't get it...what is the application to my work in instructional technology as an instructor? Is this just a generic survey for all professions? It assumes I believe poverty can be eliminated. It cannot. God says the poor will always be with us and we should care for them. Disease will not disappear. Sin is a part of the nature of man and therefore people will seek and find ways to abuse others. It cannot be eliminated but it can be controlled and minimized."* (sic)

5. *"The 13 questions on social responsibility were difficult to answer. For example I could have easily checked "does not apply" on all of them (both on the "what is" and "what should be") because I do not consider these issues (e.g. hunger, poverty, disease, etc.) as part of my profession. As a human being or on the personal level I certainly think about them and sometimes take action related to such issues (e.g., give to charity or join an environmental group) but in my "profession" which is what I think the question was asking I do not integrate these issues or include them in my teaching or research."*

6. *"The content for many questions in what is vs what should be is not appropriate for classroom instruction."*

7. *"The "What is" and "What should be" seemed unrelated to my educational technology efforts."*

8. *"I found the last series of questions confusing in terms of relating these very broad social issues to my teaching."*

While these comments are not representative of all the comments from faculty, these sorts of comments did occur often enough to provide some indication of why elements were not rated highly and why faculty may not have really seen any distinction between the elements.

In fact, it could be that the elements do represent a single construct to most participants in the study – "social *issues*" where issues are separate from forms of professional responsibility that can be codified into behavior or performance expectations.

Kaufman (2000) and Kaufman, Oakley-Browne, Watkins, & Leigh (2003) suggest that resistance to Mega is often explained by the fear that respondents have relative to understanding and accepting linkages between their professional contributions and modeling or using it.

It is important to note that the findings in this study with this population are somewhat different from findings in a similar study conducted with a practitioner population. In 2001, Guerra conducted a study to determine the perceived gaps for competencies required for certification as a Certified Performance Technologist (CPT).

Those competencies included ethics that were stated in terms taken directly from Kaufman's (2000) Mega model of societal impact.

The competencies relating to organizational impact on society received the lowest median ratings out of all the competencies for both "does apply" and "should apply."

However, the measured gap between "does apply" and "should apply" was larger than the measured gap here, and the measured perception of how often it should apply was higher than what was reported in this study.

Guerra (2001) found that general practitioners reported they should apply societal impact competencies sometimes; experts reported that these competencies should be applied frequently (p. 109-110). These ratings are higher than what faculty in the field report, and the difference is worth noting.

To explain the findings, Guerra (2001) suggests that some professionals have not yet made the *paradigm shift* (Kuhn, 1974) necessary for them "to recognize the relevance and ethical and practical importance of such competencies in their professional practice" (p.110).

Many well-known figures have continually argued that members of a profession have a professional obligation to clients and to society (Dean, 1993; Kaufman, 2000b, 2006; Westgaard, Watkins, Leigh & Kaufman, 2000).

Indeed, the findings from this study suggest that faculty, too, have not yet made the paradigm shift to seeing the relevance of social responsibility to their professional practices.

However, the difference between the findings of this study, where faculty rated Kaufman's Elements consistently low, and Guerra's (2001) study, where practitioners rated the "should apply" higher, suggest another possible explanation for the findings.

Guerra's findings indicate that practitioners are more likely to see the relevance of social responsibility to their practices than faculty teaching in graduate programs. It could be that general practitioners tend to be closer to the point of implementation of designs, and therefore are closer to the results and impact.

As a point of contrast, other participants in this study had more positive comments saying they teach about some of these elements in their courses and are concerned about both students ability to problem-solve ethical decisions and about decisions regarding the use of technology in education made at broader governmental levels.

The sample size for this study was not large enough to disaggregate the data and compare groups; however, it could be that for faculty who believe these elements are related to professional responsibilities and who explicitly integrate ethics into their courses, that the elements would load onto factors that could be interpreted according to ethical principles.

This might indicate that the persons who view social responsibility as important would also likely have a cognitive schema, or a cognitive map, of those types of responsibilities – how they are different and how they relate.

Given the results, where basically all elements were regarded as Never or Rarely applicable, the majority of participants in this study appear do not consider social responsibility (as defined by Kaufman's elements) as important in their professional activities.

However, the raw data shows that there were 7 participants in the study who rated all the elements highly (at a 5 or 6, where 5=Usually and 6=Always).

One participant rated all the elements a 6.

These seven participants would be of interest for a follow-up study investigating how they view ethics, how they teach ethics, and what their cognitive schema of ethics looks like and identify any personal or professional characteristics that would differentiate this group from those who did not accept the Mega concept as relevant to them and their curriculum.

As to the one element that did rate highly in the current study, the one reason spouse/partner abuse may have rated so highly as compared to the others is because faculty may deal with this social issue more on a day-to-day basis as a result of their advising responsibilities to adult students.

Divorce rates are anecdotally known to be high among graduate students, so faculty may be in a position where they are confronted with spouse/partner abuse (or at least interpersonal difficulties) more than other types of social issues, making them more sensitive to this particular element.

If such is the case, then the response on this item possibly indicates that faculty answered this part of the survey thinking of themselves as professionals more in terms of being a professional faculty member than a professional in the field of educational technology.

Summary

The results indicated that Kaufman's Elements of the Basic Ideal Vision did consistently measure a stable construct of social responsibility, but participants did not distinguish between different ethical principles in those elements and did not perceive the elements as anything they should be particularly committed to in their professional work.

A review of planning documents from programs corroborates this interpretation with data that is not based on self-report. No programs in educational technology in the US or Canada incorporated societal impact in their planning documents in clear, measurable terms. Some are thinking about it, but have not yet linked those to clearly stated objectives or results.

Future research possibilities include further exploration of the construct of social responsibility.

A focused study with outliers, or "experts," could provide a clearer structure or schema for social responsibility which could then be used to determine novice, intermediary and expert levels of comprehension.

Such a study could also be expanded to different populations within the same field, to other fields that deal with technology, and to the development and testing of a structural model of professional ethics.

Additionally, a change perspective that looks at "awareness" and other concerns or levels of use (Hall & Hord, 2001) might determine that the variables do not necessarily correlate because social responsibility is a new "innovation" that simply has yet to diffuse.

Bibliography

Ary, D., Jacobs, L., & Razavieh, A. (1996). *Introduction to research in education* (5th ed.). Fort Worth, TX: Harcourt Brace College Publishers.

Baumgarten, (1982). Ethics in the academic profession. *Journal of Higher Education, 53*(3), 282-295.

Berenbeim, R. E. (1987). Corporate ethics (Research report #900). New York: The Conference Board.

Brenner, S. N., & Molander, E. (1977). Is the ethics of business changing? *Harvard Business Review, 5*(1), 57-71.

Brethower, D. (2005). Yes we can: A rejoinder to Don Winiecke's rejoinder about saving the world with HPT. *Performance Improvement Journal, 44*(2): 19-25.

Caron, B., ed. (1999). *Service matters: The engaged campus.* Providence, RI: Campus Compact.

Cuban, L. (1986). *Teachers and machines: The classroom use of technology since 1920.* New York : Teachers College Press.

Cuban, L. (2003). *Oversold and underused: Computers in the classroom, 1980-2000.* Cambridge, MA: Harvard University Press.

Davies, I. (1996b). Educational technology: Archetypes, paradigms and models. In Eds. D. Ely and T. Plomp, *Classic writings on instructional technology*, pp. 15-30.

Davis, M. (1999). *Ethics and the university.* London: Routledge.

Dean, P.J. (1993). A selected review of the underpinnings of ethics for human performance technology professionals – Part One: Key ethical theories and research. *Performance Improvement Quarterly, 6*(4), 3-32.

Dean, P. J. (1999). The relevance of standards and ethics for the human performance technology profession. In H. Stolovitch and E. Keeps (Eds.), *Handbook of Human Performance Technology* (2nd ed.., pp. 698-712). San Francisco: Jossey-Bass.

Delaney, J. & Sockell, D. (1992). Do company ethics training programs make a difference? An empirical analysis. *Journal of Business Ethics, 11*, 719-727.

Dobni, D., Ritchie, J.R., & Zerbe, W. (2000). Organizational values: The inside view of service productivity. *Journal of Business Research, 47*, 91-107.

Ely, D. (1976). Creating the conditions for change. In S. Faibisoff & G. Bonn (Eds.), *Changing times: Changing libraries* (pp.150-162). Champaign, IL: University of Illinois Graduate School of Library Science. ED 183 139.

Ely, D. (1990). Conditions that facilitate the implementation of educational technology innovations. *Journal of Research on Computing in Education, 23*(2), 298-305.

Ely, D., & Plomp, T. (1996). *Classic writings on instructional technology*. Englewood, CO: Libraries Unlimited.

Finn, J. D. (1996a). A walk on the altered side. In D. Ely & T. Plomp (Eds.), *Classic writings on instructional technology* (pp. 47-56). Englewood, CO: Libraries Unlimited, Inc.

Finn, J. D. (1996b). Professionalizing the audio-visual field. In Eds. D. Ely and T. Plomp, *Classic writings on instructional technology* (pp. 231-241). Englewood, CO: Libraries Unlimited, Inc.

Fitzgerald, M., Orey, M., & Branch, R. (2003). *Educational Media and Technology Yearbook 2003* (Volume 27). Westport, CT: Libraries Unlimited.

Fitzgerald, M., Orey, M., & Branch, R. (2004). *Educational Media and Technology Yearbook 2004* (Volume 28). Westport, CT: Libraries Unlimited.

Ford, R., & Richardson, W. (1994). Ethical decision making: A review of the empirical literature. *Journal of Business Ethics, 13*, 205-221.

Frankel, M. (1989). Professional codes: Why, how, and with what impact? *Journal of Business Ethics, 8*, 109-115.

Gall, M., Gall, J., & Borg, W. (2003). *Educational research: An introduction* (7th ed.). Boston: Allyn and Bacon.

Guerra, I. (2001). *A study to identify key competencies for performance improvement professionals.* Unpublished doctoral dissertation. Florida State University.

Hall, G., & Hord, S. (2001). *Implementing change: Patterns, principles and potholes.* Boston: Allyn & Bacon.

Healy, J. (1990). *Endangered minds: Why our children don't think.* New York: Simon & Schuster.

Healy, J. (1999). *Failure to connect: How computers affect our children's minds – and what we can do about it.* New York: Simon & Schuster.

Kaufman, R. (1996). Needs assessment: Internal and external. In D. Ely & T. Plomp (Eds.), *Classic writings on instructional technology* (pp. 111-118). Englewood, CO: Libraries Unlimited, Inc.

Kaufman, R. (1997). A new reality for organizational success: Two bottom lines. *Performance Improvement, 38*(8), 3.

Kaufman, R. (2000a). Education past, present and future: From how to what to why. *International Journal of Educational Reform, 9*(1), 2-8.

Kaufman, R. (2000b). *Mega planning: Practical tools for organizational success.* Thousand Oaks, CA: Sage Publications.

Kaufman, R. (2006). *Changes, choices and consequences: A guide to Mega thinking and planning.* Amherst, MA: HRD Press.

Kaufman, R., Corrigan, R., & Johnson, D. (1969). Towards educational responsiveness to society's needs: A tentative utility model. *Socio-Economic Planning Sciences, 3*, 151-157.

Kaufman, R., Oakley-Browne, H., Watkins, R., & Leigh, D. (2003). System (and systems) thinking. In Ed. R. Kaufman, *Strategic Planning for Success: Aligning people, performance, and payoffs.* San Francisco: Jossey-Bass/Pfeiffer.

Kuhn, T. (1974). *The structure of scientific revolutions* (2nd ed.). Chicago: University of Chicago Press.

Matchett, N. (2008). Ethics across the curriculum. *New Directions for Higher Education: Practical Approaches to Ethics for Colleges and Universities* (ed. S. Moore): *142*, 25-38.

Petrick, J. & Quinn, J. (1997). *Management ethics: Integrity at work.* Thousand Oaks, CA: Sage Publications.

Procario-Foley, E. & Bean, D. (2002). Institutions of higher education: Cornerstones in building ethical organizations. *Teaching Business Ethics, 6*, 101-116.

Verschoor, C. (1998). Corporations' financial performance and its commitment to ethics. *Journal of Business Ethics, 17*, 1509-1516.

Stephanie Moore, PhD., CPT

Stephanie L. Moore is Director of Engineering Instructional design at the University of Virginia for the School of Engineering and Applied Science where she works on distance education and will be teaching ethics for undergraduate engineers.